Welcome to the San Diego Zoo

The San Diego Zoo takes visitors on a walking adventure through botanical canyons and lush rain forests, past rare and endangered animals and exotic plants. Here, giant pandas eat bundles of bamboo, Queensland koalas nestle into eucalyptus trees, polar bears swim through a tundra habitat, and one of the nation's largest collections of birds flies through numerous aviaries.

A vast assortment of mammals, birds, reptiles, and amphibians representing over 800 species and subspecies lives in the San Diego Zoo's canyons and mesas. We hope you enjoy discovering these wonders of wildlife as much as we enjoy sharing them with you. As our founder, Dr. Harry Wegeforth, said, "A zoo is just about the most fascinating place in the world."

SAN DIEGO ZOO

OFFICIAL GUIDEBOOK

Contents

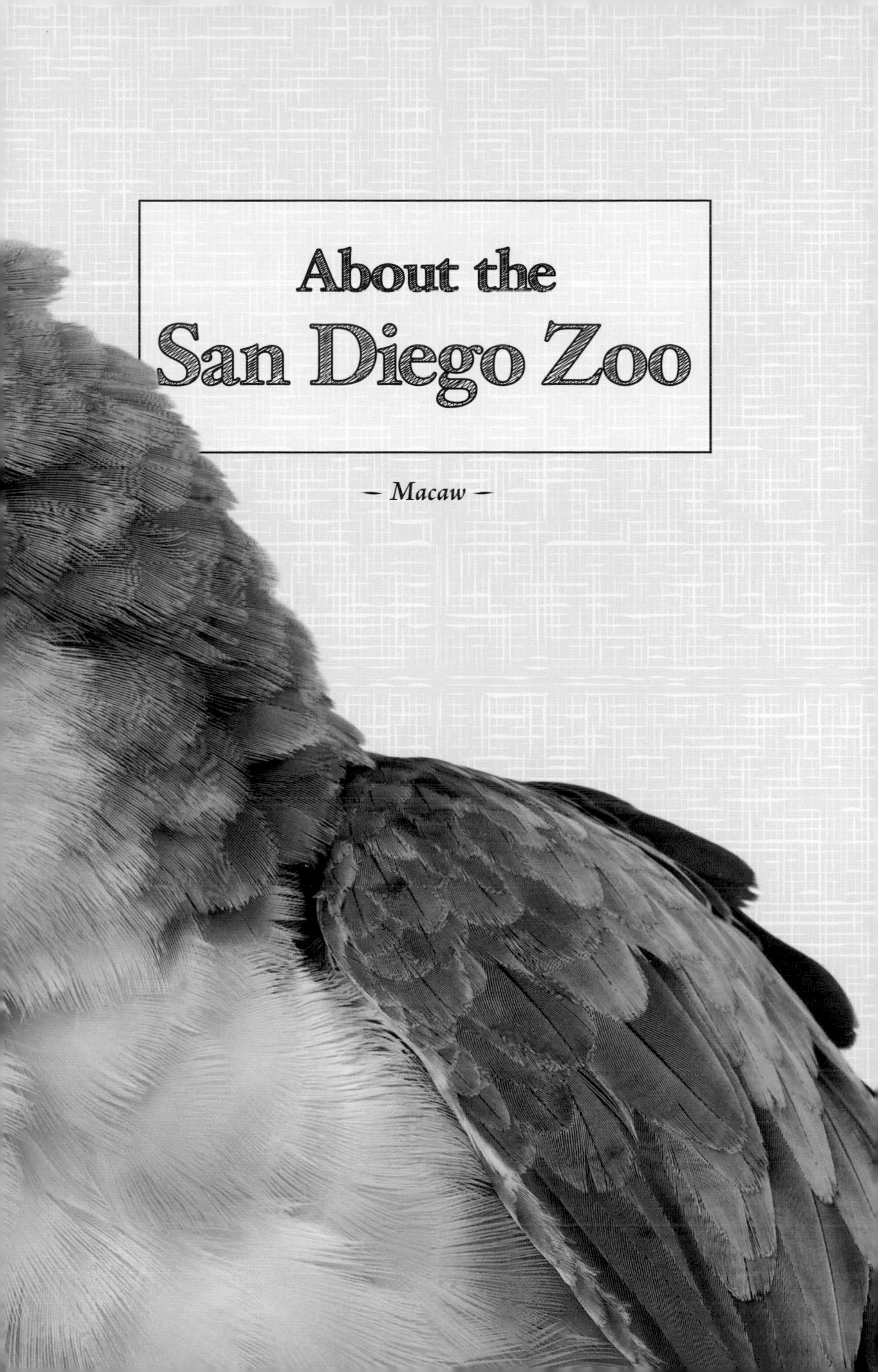

About the San Diego Zoo

– Macaw –

A lion's roar can be heard up to three miles away. Males can roar when they are about one year old, and females can roar a few months later. Lions also communicate through growls, snarls, hisses, meows, grunts, and even puffing.

About the San Diego Zoo

It all began with a roar. In 1916, Dr. Harry M. Wegeforth heard a lion at the Panama-California International Exposition in Balboa Park and asked, "Wouldn't it be wonderful to have a zoo in San Diego? I believe I'll build one."

"Dr. Harry" established the Zoological Society of San Diego on October 2, 1916, with a small collection of animals that had been featured in the International Exposition. Today, the Zoo cares for more than 4,000 animals and 700,000 exotic plants. Through its umbrella organization, San Diego Zoo Global, the Zoo and its other facilities breed more than 165 endangered species, have reintroduced 33 species into the wild, and participated in more than 100 conservation field projects in over 35 countries. Built on 100 acres, the Zoo is dedicated to the conservation of endangered animals and their habitats.

San Diego surgeon and Zoo founder Dr. Harry Wegeforth took a hands-on approach from the very beginning.

San Diego Zoo Milestones

1916 The San Diego Zoo originally consisted of rows of cages that had been part of the Panama-California International Exposition. Dr. Harry Wegeforth formed the Zoological Society of San Diego to fund caring for the animals.

Circa mid-1920s New living spaces were built for the growing collection. The San Diego Zoo was the first in the United States to create exhibits featuring large moats with a sloped landscape, which meant there were no bars between the animals and the guests.

1925 Snugglepot and Cuddlepie, the Zoo's first koalas, were first for another reason: they were the only koalas outside of Australia at the time. The pair was a gift "from the children of Sydney, Australia, to the children of San Diego."

1927 The Zoo quickly became a part of the education of San Diego schoolchildren, seen here on a field trip to learn about animals. In 1939, the Zoo began a collaborative education program with San Diego schools that continues today: all second grade students visit the Zoo to enhance curriculum learning about life cycles.

1931 Scientists from every corner of the globe came to observe two thriving five-year-old mountain gorillas named Mbongo and Ngagi at the Zoo. At the time, there were many myths but few facts about gorillas, making every detail of their care a learning experience. Bronze busts of them are just inside the Zoo entrance.

Circa late-1940s In the post–World War II baby-boom years, the Zoo became even more popular. San Diego's population had grown, and servicemen that had been based in San Diego brought their families from across the nation.

1957 The Children's Zoo, a "zoo within a zoo," was a pioneering concept that scaled exhibits to a child's-eye view. You can explore the kid-centric exhibits in the Discovery Outpost zone.

1955 Lights, camera, action (plenty of it!): the television show *Zoorama*, broadcast live from the San Diego Zoo, began production. Originally created for the San Diego market, the show was picked up for national broadcast and later sold as a syndicated series.

1960 The first koala was born in North America. This was the beginning of a successful breeding program that allowed the Zoo to share koalas with other zoos. The Zoo has loaned koalas to over 65 institutions in 12 countries.

1972 Originally conceived as a preserve dedicated to breeding exotic animals, the San Diego Wild Animal Park—known today as the San Diego Zoo Safari Park—has had much success in helping build animal populations.

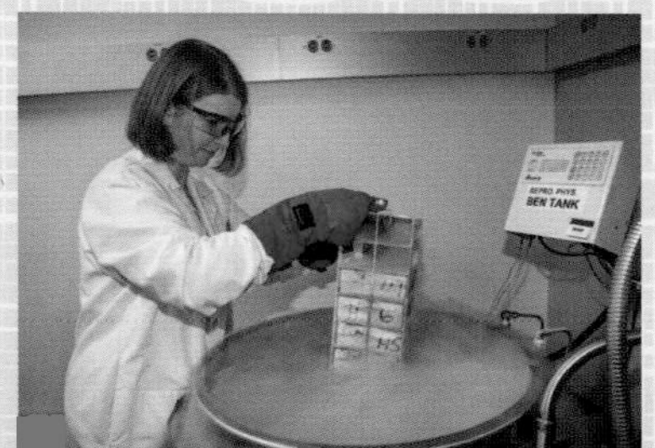

1975 San Diego Zoo scientists establish the Frozen Zoo. This cryogenic storage of tissue and DNA of endangered and extinct wildlife is meant to provide scientists of the future with information about animals of today and yesterday.

1986 A rebuilding of the Zoo began, using the concept of "bio-climatic" zones to immerse guests in the plants, sounds, and wildlife of specific climates and habitats found around the world.

1994 Karen, a two-year-old Sumatran orangutan, was diagnosed with a hole in her heart and became the first of her species to undergo open-heart surgery. The devotion and expertise of more than 100 medical personnel at UCSD Medical Center as well as the Zoo made the treatment a success. Karen is still thriving today.

1996 Giant pandas Bai Yun and Shi Shi arrived at the San Diego Zoo from the People's Republic of China as part of a 12-year research loan that has been extended multiple times. In 1999, their offspring Hua Mei was born. She is the first US-born panda cub to survive and is now a grandmother herself!

Tips and Information

The San Diego Zoo is open every day of the year, rain or shine, including holidays. Parking is free.

Tickets for the bus tour are sold at the corner of Front Street and Treetops Way, near the Zoo's entrance. The tour goes through Tiger River, Ituri Forest, Panda Canyon, Bear Canyon, and the outskirts of Elephant Odyssey.

In order to fully experience the animals, plants, and various shows and presentations offered throughout the day, the Zoo recommends a minimum visit time of at least three to four hours.

The Zoo is built on 100 acres of lush canyons, with many landscaped pathways leading through bioclimatic zones and into aviaries. There are winding paths, hills, and valleys—with lots of places to stop and relax or grab a bite to eat—so a good pair of walking shoes is recommended.

There is a 35-minute Guided Bus Tour that travels through many sections of the Zoo, giving guests a different view of the animals than they can see on foot. There is also an Express Bus that stops at several locations to help guests access different areas.

For guests with difficulty walking, there are a variety of options. The Zoo offers an "Easy Access Pass" that allows these guests and their parties to go to the head of the line at the Bus Depot and Skyfari aerial tram. Complimentary shuttles are available to assist mobility-challenged visitors in negotiating some of the most steeply graded areas. Guest Services provides an ADA packet with information on these services as well as descriptions of exhibits and show access.

CATCH A GLIMPSE

In general, the best time to visit the Zoo is first thing in the morning, when many animals are most active. Some animals, however, can be difficult to see. The exhibits mimic the animals' natural environment and give them plenty of room to hide. For many species, this ability to hide is important to their well-being.

The Zoo and Safari Park have been home to more than 100 Bengal, Siberian, Sumatran, and Malayan tigers. The Zoo is currently home to Malayan tigers.

Every year during the month of October, to celebrate the Zoo's founding, children ages three to 11 are admitted for free (accompanied by a paid adult). The Guided Bus Tour and Skyfari aerial tram must be purchased separately.

Strollers, lockers, bag-check, wheelchairs, and electric scooters are available for rent on a first-come, first-served basis near the entrance.

ATMs are located throughout the Zoo and outside the entrance.

There are picnic areas located outside the Zoo's entrance. Guests may bring personal food items into the Zoo, but glass, alcohol, and straws are not allowed.

A staffed first-aid station is located on the north side of the Zoo's Reptile House. It is open daily during regular Zoo hours.

The Zoo is a quarantined facility. State and federal regulations do not allow for pets to accompany guests into the Zoo. There are no kennels for pets on Zoo grounds.

There is no smoking allowed on the grounds of the San Diego Zoo.

The top deck of the guided tour bus is popular with many out-of-state visitors, but the views are equally as good on the protected lower deck.

Unique Activities

The Zoo offers a variety of unusual attractions throughout its 100 acres. One of the most iconic is the **Skyfari aerial tram.** Opened in 1969, the tram provides a bird's-eye view of the Zoo, Balboa Park, and downtown San Diego. The Skyfari has carried almost 70 million riders since it opened, with more than half of all Zoo-goers riding the Skyfari at least one time during their visit. It's free to members, but visitors with regular tickets to the Zoo must pay a separate fee.

Across from the Skyfari's east entrance is the Zoo's **Reptile House**, which features one of the largest colonies of Galápagos tortoises in the world. Ten of the tortoises are well over 100 years old! When a keeper is present, guests may be invited to feed vegetable snacks to the tortoises and scratch their neck. Some of the tortoises are shy, while others love the attention.

In the interaction area, those guests offering keeper-provided snacks or wearing the colors found in flowers—a favorite food—often get the tortoises' attention.

During the summer, the Zoo stays open later; bands and performers add to the fun after dark.

At the southwest edge of Discovery Outpost is the **Insect House**. Tucked around the corner from the children's Petting Paddock, the Insect House is easy to miss but it's well worth the visit. Inside the sliding doors are both common household insects like ants and bees—and unusual ones like giant roaches and millipedes. There's often a staff member on hand to answer questions, too.

At the other end of Front Street is **Australian Outback**, which features the largest Queensland koala breeding colony outside of Australia. Since the males can be territorial, they each have their own perch; the more social females and their joeys share a large space. The exhibit, which opened in 2013, includes winding paths through eucalyptus groves, 7 species of marsupials, and 25 species of birds.

The Zoo received its first two koalas in 1925. Today, Australian Outback is home to 7 kinds of marsupials and about 25 species of birds. The Koala Cam, which can be viewed online, spotlights the largest enclosure, where many females and their joeys live.

Eleven pandas have lived at the Zoo since the first pair visited on a 100-day loan from China in 1987. The Zoo's adult female, Bai Yun, gave birth to her sixth cub in 2012.

Down Park Way is **Panda Canyon.** Like the Outback's ties to Australia, Panda Canyon has important links to the international community. In 1996, the Zoo received two giant pandas on long-term loan from China. Since then, millions of people have visited the Zoo's pandas. In 2012, the area underwent renovations. Today, the area includes larger exhibit areas, extra vegetation and climbing structures, and elevated viewing paths.

In addition to its rare and endangered animals, the Zoo is well known for its horticultural collection. Nestled between Center Street and Treetops Way is **Fern Canyon**, one of the Zoo's hidden horticultural jewels. Here, a shaded pathway winds through a rain forest refuge of lush ferns, lily vines, purple-flowering jacaranda trees, and giant Burmese honeysuckle. The Zoo offers self-guided plant maps to Fern Canyon and other botanical areas just past the entrance turnstiles.

A range of ferns, from small sword ferns to 40-foot-high tree ferns inhabit Fern Canyon. During spring and summer, the garden is covered by a lush purple canopy of blooming jacaranda.

BEHIND THE SCENES

Each year, the Nutritional Services Department buys 10.5 million crickets, 288,000 mice, 20,600 pounds of femur bones, and 115,000 pounds of frozen fish for the Zoo's meat eaters. The vegetarians receive, among other things, 43,200 heads of lettuce, 1,000 cases of apples, 17,600 pounds of bananas, 37,500 pounds of carrots, and 37,000 pounds of yams.

Special Experiences

– Giraffe –

Special Experiences tours enable guests to meet animal ambassadors, such as giraffes, cheetahs, anteaters, and more.

Special Experiences

THE SAN DIEGO ZOO OFFERS A VARIETY OF TOURS and animal interactions. Guests can meet various animal ambassadors up close like elusive leopards and endangered macaws, learn about the Zoo's conservation work, see what goes on behind the scenes, or explore the Zoo's world-class botanical garden. These Special Experiences are separate from the cost of admission and are subject to availability. In addition, there are numerous educational programs designed to extend the guest experience and provide an in-depth understanding of the Zoo's plant and animal collections and its worldwide conservation efforts. These programs include expert lectures, weeklong summer camps, photography classes, sleepovers, and other fun and educational activities for kids and families.

The Zoo's macaw ambassadors participate in various special events.

Backstage Pass

Backstage Pass allows guests to meet, touch, and help train some amazing animal ambassadors. The encounter begins as a professional photographer takes a picture of each family/group with one of the Zoo's animal stars. Guests can also get a photo with a featured animal, such as a cheetah, and visit the rhino exhibit, where they can touch and feed the animals. Zoo hosts provide insider stories about the staff and animal residents, as well as information about the Zoo's conservation efforts. The Zoo's photographers take pictures during the tour, and a CD of images is available for purchase later that day. The 90-minute experience includes unlimited iced tea, soda, or water, and the ticket allows guests to spend the rest of the day at the Zoo enjoying special perks like discounts on merchandise in the gift shops. Animal access is subject to change based on the welfare of the animals.

DID YOU KNOW?

There are more than 16,000 plants and animals on the Endangered Species List, including the clouded leopard. Because of habitat loss, disease, and the illegal wildlife trade, tigers, elephants, giant pandas, and many more species are also at risk of extinction.

Inside Look Tour

During the Inside Look Tour, guests can find out how the Zoo works to accomplish its goals of wildlife conservation and education. Guests tour the Zoo in a small cart and visit off-exhibit areas, where they can see how the Zoo takes care of its animals. This two-hour tour is offered daily at various times.

Mission Conservation

Conservation takes a great deal of cooperation. During the Inside Look Tour, guests learn about San Diego Zoo Global's role in the Species Survival Plan (SSP) for various animals, such the Micronesian kingfisher. The SSP enables zoological facilities to work together to make decisions about how to best manage a captive population to meet conservation objectives such as captive breeding. Many rare or endangered species have an SSP. Animal exchanges between zoos are often done based on the animal's SSP. In many ways, the program is like a "dating service."

Discovery Tour

The Discovery Tour takes place in an expedition shuttle cart and includes areas of the Zoo not seen from the regular bus tour. Here, a guide shares stories about the Zoo's plants and animals and explains how the Zoo's conservation work helps wildlife. Guests can step off for better views of their favorite animals, take photographs, and ask as many questions as they'd like. The tour includes different options like What's Growin' On, a botanical tour, and Ape Adventure, which focuses on gorillas, siamangs, bonobos, and orangutans. During each 60-minute tour, guests can see 70 percent of the Zoo—an ideal option for people who are only visiting for a short time.

FIELD NOTES

During Ape Adventure Discovery Tour, guests learn about the ape species in Lost Forest. Orangutans, for instance, don't live in troops with a strict social structure. Dominant males have established territories that females can enter—and leave—at any time. Orangutans are also very intelligent. When they're given an enrichment puzzle, they don't manipulate it right away. They wait to come up with a solution before applying it.

Exclusive VIP Experience

The Exclusive VIP Experience involves a team of professionals who customize each guest's visit to suit his or her needs and interests. The tour provides access to off-exhibit areas and allows for plenty of animal interaction (if desired). During this tour, guests spend up to either five or eight hours exploring the Zoo with their own tour guide. Each Exclusive VIP Experience includes a meal. Animal presentations and behind-the-scenes areas are subject to change based on the health and welfare of the animals.

Aloes, succulent members of the lily family, are mostly native to Africa. The thick leaves store water and are often eaten by animals in times of drought—if they can get past the sharp spines!

Horticulture Tour

The Horticulture Tour shows guests a different side of the San Diego Zoo—its green side. A Horticulture Department staff member provides a private tour that is custom designed to fit the interests or needs of the group. For instance, the tour might provide a general overview of the Zoo's gardens and history, delve into a certain collection of plants, such as the Zoo's ficus or palm trees, highlight a specific garden area like Fern Canyon or the orchid greenhouse, or focus on growing the types of bamboo and eucalyptus that are fed to the pandas and koalas. The group must be at least 15 members, making it ideal for a garden club or school group. Horticulture Tours are one and a half or two hours, depending on preference.

DID YOU KNOW?

San Diego's mild Mediterranean climate makes it possible to grow more different kinds of plants than almost anywhere else in the United States.

Education Programs

With more than 3.5 million visitors each year, the Zoo offers a wide variety of supplemental educational programs. For families with three- to six-year-olds, the monthly KinderNights features snacks, activities, and interactions with an animal ambassador. Younger kids 18 months to three years can attend KinderTots with an adult on select Saturdays. For children entering kindergarten through seventh grade, the Zoo's camps provide day- or week-long animal adventures. All camps feature animal encounters, behind-the-scenes visits, bus rides, games, edible food crafts, and midday shows. Kids who want to explore their inner artist can attend wildlife art camps taught by professional art teachers. And younger kids ages four and five can participate in Cub Camp, a modified half-day program. Older students entering 8th to 12th grade can attend Survivor Challenge, an action-based camp that takes them directly into a keeper's or animal's world. Teens can also take part in Zoo Corps, a volunteer program that sets up discovery stations throughout the Zoo, teaching them about animal diets, wildlife conservation, animal enrichment, and endangered species.

For adults, an after-hours program, Twilight Trek, highlights the Zoo's nocturnal animals and provides stories from educators and other inside information. The theme varies each month, and dessert and hot drinks are served. For early birds, the Sunrise Surprise Strolls start at 7:30 on select Saturday and Sunday mornings. During these walks, experienced Zoo educators provide updates on the Zoo's animal collection. There are also several photography programs, including one where guests visit various locations around the Zoo before and after hours, and a two-part class that can be tailored for any skill level.

Toucans may look heavy, but their large bill is remarkably lightweight, enabling them to perch easily on the smallest of branches (and guests!). Their bill is made of keratin—the same material found in human hair and fingernails.

During summer and winter camps, school-age children learn from Zoo educators, meet animal ambassadors, and go behind the scenes of certain exhibits.

Giant pandas grasp bamboo stalks using their five digits and a special bone that extends from their wrist called a "pseudo-thumb." Then they use their teeth to peel off the tough outer layers to reveal the soft inner tissue of the stalk.

Early Morning with Pandas

Truly an "early bird special," the Early Morning with Pandas tour allows participants to be the first to say "good morning" to the Zoo's popular giant pandas. This two-hour adventure begins before the Zoo opens. While the black-and-white bears munch their bamboo breakfast, guests listen to fun stories about their unique personalities, as well as all the interesting facts a panda-fan would want to know.

The tour continues with a drive around the Zoo in a comfortable cart, with stops to enter some behind-the-scenes areas to discover firsthand the varied ways the Zoo meets its animals' diverse needs. And while the giant pandas are the stars, guests are able to get a special look at other wildlife along the way, experiences that can include anything from up-close encounters with camels to a special view of gorillas and other primates. Many of the animals are more active—and vocal—in the morning, and guests truly enjoy hearing their "calls of the wild."

CATCH A GLIMPSE

One of the stops guests might make on this tour is to visit the bonobos' habitat on Hippo Trail. Earlier scientists, who thought bonobos were just a smaller version of the common chimpanzee, called them "pygmy chimpanzees." But bonobos are more slender with smaller skulls and ears. They look quite distinguished with their "hair" neatly parted down the middle of their head.

The bonobos' social structure is unique and complex: in their largely peaceful society, the females rule the roost.

Front Street

– Flamingo –

In Chinese culture, Mandarin ducks represent fidelity and love since the males do not abandon their mate and ducklings as some other bird species do.

Front Street

FRONT STREET IS A WHIRLWIND OF ACTIVITY, WITH music, shows, cafés, shops, face painting, and life-sized statues for photo ops. Here, peacocks parade around the plaza, calling to one another and often spreading their stunning, famous tail in displays that impress other peafowl—and Zoo visitors! Exotic macaws zoom overhead in a free-flight demonstration at the start of each day to "officially" open the Zoo. And from the main stage, keepers educate guests on conservation with the help of animal ambassadors like a blue-tongued skink, a harpy eagle, and a cheetah and its domestic companion dog.

Front Street stretches from Discovery Outpost and the Children's Zoo all the way to Australian Outback and Elephant Odyssey. The Reptile House is also located off Front Street, as are the Skyfari and Guided Bus Tour.

When unfolded, a peacock's tail feathers can be six to seven feet wide.

Flamingo

The pink color of flamingos' feathers comes from carotenoid pigments in the algae and small crustaceans that they eat. At the Zoo, the flamingos are fed a special pellet diet that includes this pigment to help them maintain their beautiful color. Unlike many animals that rely on their coloring to help them blend into their surroundings, flamingos really stand out. They tend to live in places where the lagoons are pretty bare of vegetation, so few other birds or animals come there, making them less attractive to predators.

Flamingos have distinctive eating habits. Using their long legs, they wade into deep water, submerge their head, and with their beak upside down, draw in water and mud at the front of their bill, and pump it out at the sides. Fringed plates in their mouth called lamellae act like filters, trapping shrimp and other small water creatures that the birds eat.

Flamingos build their nest out of mud, using their hooked beak to dig and deposit wet sediment in a pile. Their nest can reach about 12 inches high—tall enough to keep their egg dry if the water rises. Flamingos lay a single large egg,

BEHIND THE SCENES

Each year during breeding season, the keepers build "starter" nests in rows for the flamingos. Then they draw a map of the flamingos' island, marking the nests to keep track of which pair of birds has claimed which nest and when each egg has been laid. This helps the Zoo anticipate when the eggs will hatch.

Flamingos breed only when the conditions are just right. Any change to the flock or exhibit—major or minor—can start or stop breeding.

which is incubated by both parents. Newly hatched chicks have gray downy feathers, a straight, pale beak, and swollen pink legs, both of which turn black within a week. After hatching, they stay in the nest for 5 to 12 days. During this time, they are fed "crop milk," a fat- and protein-rich secretion from their parents' upper digestive tract. Both males and females can feed chicks this way, and other flamingos can act as foster feeders. Researchers believe the begging calls of hungry chicks help to stimulate the secretion of crop milk.

Crested Screamer

FIELD NOTES

Although they forage on the ground, crested screamers fly straight up into the trees when they feel threatened by predators. Once safely out of reach, they do what they do best—call out loud and long to warn others. In their native range, large groups of crested screamers gather and roost in shallow water for the night, often producing a deafening chorus of strident calls.

Crested screamers are classified in the same scientific order as waterfowl (ducks) but are in their own unique family. These large, gray birds look like turkeys or raptors on stilts. Native to South America, screamers are excellent swimmers even though they have very little webbing on their feet. They're also great flyers, but they're nonmigratory birds.

Crested screamers were given their name for good reason: their loud calls are among the noisiest in the bird kingdom! Their shrieking vocalizations discourage many predators. They also have a sharp, keratinous spur on each wing to help them protect themselves.

Screamer chicks are precocial, which means that upon hatching their eyes are open, they are covered with down, and they are mobile. Adult screamers are herbivores, but chicks require a diet higher in protein. Both parents care for the chicks, keeping them close and brooding them when they're cold. Chicks are fully independent by the time they're 14 weeks old.

Crested screamers stay in the same area all year, near fresh water, spending most of their time feeding on seeds and the leaves of succulent plants.

Cycad

With 104 species and many mature specimens, the San Diego Zoo has one of the most extensive cycad collections in the world, including 29 species that are conservation worthy. Cycads are truly prehistoric plants, dating to the late Paleozoic era. In fact, they were already widespread by the time dinosaurs ruled the world, providing food for many of the giant plant eaters. Cycads managed to adapt and survive through the world's drastic climate changes to become even more common in our era. Today, the young leaves and seeds of many cycad species continue to be an important food source for many animals. Interestingly, most cycads are poisonous to humans unless they have been processed to remove their natural alkaloids. When cooked, however, the starch from their stems and seeds is not only edible but has been an important food source for humans in times of famine.

Although many cycads resemble palms, they're actually more closely related to conifers. Yet unlike conifers, which usually have male and female cones on the same plant, cycads often have the male and female cones on separate plants. Some cycads are wind pollinated, but most are pollinated by insects, birds, and animals such as squirrels or coatimundis.

CATCH A GLIMPSE

As you walk from the north exit of the Sandwich Company, toward the Australian Outback, and down into Center Street, take time to contemplate the prehistoric cycads. The plants here represent each of their continents of origin: Asia, the Americas, Africa, and Australia.

Ficus

Ficus trees are considered a cornerstone species of a healthy forest. Virtually every living thing in the jungle depends on them in some way. Ficus trees provide safety, shelter, and nesting material for a wide variety of animals throughout different regions. They also provide food for tree dwellers like monkeys, birds, and fruit bats, which eat their sweet fruit (figs) and disperse the seeds.

Native to south and southeast Asia and Australia, the *Ficus* genus has hundreds of species and many notable members, including the Bo tree *Ficus religiosa* that shaded Buddha, the rubber tree *Ficus elastica* from India, and the huge Moreton Bay trees *Ficus macrophylla* and *Ficus watkinsiana* that are popular with visitors to the Zoo and Balboa Park. One of the Zoo's Moreton Bay figs has a trunk circumference of more than 38 feet. And the towering *Ficus watkinsiana* behind the flamingo lagoon is the largest member of its species in the country.

Many people marvel at what look like the multiple trunks of the Zoo's ficus trees, sometimes thinking that the Zoo planted many trees in one spot that have grown together. In fact, they're seeing one individual tree. The structures growing from the underside of the tree's branches are aerial roots. The tree sprouts aerial roots to obtain extra moisture from the humid air. These roots start out as bumps on the branch and grow downward. When they hit soil, they continue growing down and begin to thicken. After many years, these roots look and function like extra trunks, supporting the tree's branches.

Behind The Scenes

The Zoo's ficus trees supply about 2.5 tons of browse each week for the apes, elephants, rhinos, tapirs, and other animals. For some animals, ficus browse is crucial. Because the Zoo has a good supply of the trees, it shipped fresh ficus browse to the Cincinnati Zoo for their rare Sumatran rhinoceroses for more than 20 years. The ficus care packages had a positive effect on the endangered species' overall health and well-being.

Peafowl

Although most people call these birds "peacocks," the word really only refers to the males. Female peafowl are called peahens, and babies are called peachicks. There are three species of peafowl—Indian, green, and Congo—but the most common are Indian peafowl, found in many zoos and parks.

Indian peacocks have flashy plumage, with a bright blue head and neck and long, spectacular tail feathers that are designed to attract a mate. Peacocks grow their first train by the time they're two years old. Their train gets longer and more elaborate every year, reaching its maximum splendor by the time they're five or six years old.

FIELD NOTES

Peafowl have a daily routine. To avoid predators, they roost overnight in large groups in tall, open trees. In the morning, they break into small groups, foraging on the ground for grain, insects, small animals, berries, figs, leaves, and seeds. At midday, they drink, preen, and rest in the shade. Then they forage again for food, take one last drink, and return to their roost for the night.

Peahens are a mottled brown. Their brown feathers help them blend in with the vegetation, preventing predators from seeing them while they are incubating their eggs. While peahens may not look as impressive, they have a big job: nesting and raising their chicks alone. After breeding, they make a scrape in the ground, line it with sticks, and lay several light green or tan eggs. Peahens sit on their eggs almost constantly for about four weeks.

Peachicks can walk and forage right after they hatch, but they are very vulnerable. It takes two weeks before they can flap up into a tree for safety, where they huddle under their mother's wings. By two months, they look just like their mother (though half her size). Out of every six chicks that hatch, usually only two survive to join the rest of the flock.

Urban Jungle

— Rhinoceros —

Giraffes can run up to 35 miles per hour for short distances to escape their predators—mainly lions and crocodiles.

Urban Jungle

It's the only place in the busy city of San Diego where guests can get close to long-legged giraffes, see an endangered greater one-horned rhino, try to catch a glimpse of an elusive clouded leopard, and watch animals like cheetahs and domestic dogs play together in a unique space. The largest exhibit in Urban Jungle is home to a herd of Masai giraffes, which live with smaller Soemmerring's gazelles. Here, low feeding stations line the front of the exhibit so guests can watch the giraffes eat their herbivore pellets with their long, dark tongue.

Urban Jungle is also an interactive zone where the special Backstage Pass program takes place. In this program, guests have the chance to interact with a greater one-horned rhino and meet some of the Zoo's animal ambassadors up close.

Native to the rain forests of Southeast Asia, binturongs walk like bears but dart up and down trees much like cats.

Giraffe

The word *giraffe* is derived from the Arabic *zarafah,* meaning "one who walks swiftly." Giraffes are the only mammals born with horns. Their horns are comprised of soft cartilage when they are born and harden as they mature. Reaching up to 18 feet, giraffes are the tallest living land mammals. They range from 2,000 to 3,000 pounds.

Calves can walk within an hour of birth. When they get a little older, they naturally form a group called a crèche—a type of kindergarten where they develop physical and social skills through play under the watchful eye of a designated guardian (usually one of the mothers or grandmothers). While the mothers move about feeding, grooming, and socializing, the calves remain clustered together. The youngsters explore their surroundings, taking frequent breaks to check in with their mother for nursing sessions or to munch on hay or grain pellets.

DID YOU KNOW?

Just like humans, giraffes have seven neck vertebrae. Each giraffe's vertebra, however, can be over 10 inches long.

Greater One-Horned Rhino

Also known as Indian rhinos, greater one-horned Asian rhinos are native to swampy areas of northeast India and Nepal. They have a large head, broad chest, thick legs, poor eyesight, excellent hearing—and a fondness for rolling in the mud. Although they look armor-plated, they are actually covered with a layer of skin that has many folds. Like all rhinos, greater one-horned Asian rhinos are herbivores, eating grasses or leaves, depending on the species. At the San Diego Zoo, they are fed hay, browse, and high-fiber biscuits.

For ages, rhino horns have been used in folk medicine to treat illnesses, especially fevers. Yet rhino horns are made of keratin and have no healing properties. In some countries, rhinos are being dehorned, a process that removes the valuable horn but leaves the animal alive and well. This minimizes poaching. Efforts to protect these endangered animals have been successful. At the beginning of the 20th century, there were fewer than 200 greater one-horned rhinos in India and Nepal; today, with the help of the International Rhino Foundation, the population has grown to more than 2,800.

Mission Conservation

San Diego Zoo Global provides support and funding for rhino conservation efforts in India, Nepal, and Sumatra. One of its conservation partners, the International Rhino Foundation, is working to increase the greater one-horned rhinos' population in India and Nepal to 3,000 by the year 2020. More than 60 of these rhinos have been born at the Zoo's other conservation facility, the San Diego Zoo Safari Park, since 1975. One of these calves is fifth generation, the first such baby born in any zoo or wildlife park.

A wolf's howl is a celebration, whether by an individual or a group. They also whine, growl, bark, whimper, and squeak!

Behind the Scenes

The San Diego Zoo's first animal ambassador, a cheetah named Bong, was a gift from a family of animal adventurers. The family had moved from Africa to New York and cared for Bong there, even walking him in Central Park. They brought Bong to San Diego in 1933. He lived at the Zoo for 11 years, accompanying the Zoo director on speaking engagements and education events.

Animal Ambassador Play Area

This large area in Urban Jungle allows visitors to see a variety of the Zoo's animal ambassadors, which normally live in off-exhibit areas. Trained to travel and be compatible, these ambassadors and their trainers educate guests about various species and provide conservation news. When they're not out with their trainers, the animals may spend time in the Animal Ambassador Play Area.

The play area includes a pond with an overhanging rock that allows the animals to jump, splash, and cool down. Zoo guests are often surprised at the unusual pairings they see there. For example, at different times, there might be a cheetah playing ball with a domestic dog, a small New Guinea singing dog chasing a large arctic wolf, or a zebra hanging out with her Sicilian donkey companion. Different animals have their own way of making the most of the area: some like to rub against the trees or wall, while others can't get enough of splashing in the pond.

The animals are transported to the exhibit in various ways. Some are walked over from Wegeforth Bowl, others arrive in an air-conditioned van, and others travel in a custom cart or a shaded golf cart. The animal ambassadors are then walked into the yard on a leash. Once inside, the collar and leash come off and the fun begins! Guests might even have an opportunity to speak with one of the trainers about the animals in the exhibit as they are coming or going; it's a great way to learn about each animal and its partner.

Clouded Leopard

FIELD NOTES

Clouded leopards climb upside down on branches and hang by their back feet, using their front paws to snatch prey. They ambush their prey from the treetops, landing on their target's back and delivering a deadly bite. Despite their small size, they can take down large hoofed stock this way.

Few people have seen these elusive cats, either in their habitat in Southeast Asia or in zoos. Officially recorded as a species in 1821, clouded leopards remain just as mysterious today as they were nearly 200 years ago. Most of what researchers know about them comes from observing them in zoos.

Named for their cloud-like spots, these three-foot-long cats are most closely related to snow leopards and are in the same taxonomic subfamily as tigers, lions, jaguars, and true leopard species. Within the cat family, they're distinct for having the largest gender size difference, with males reaching nearly twice the size of females, and the longest tail and canine teeth of any cat in relation to body size. For instance, their two-inch-long canine teeth are the same size as a tiger's, even though tigers are 10 times larger! Their jaws can also open wider than any other cat's. And their oblong pupils are unique among cats, never dilating to a fully round shape as big cats' do, yet never shrinking to vertical slits as happens in small cats.

With fewer than 300 clouded leopards in zoos, these cats are rare. They're also in danger of extinction in the wild. Although protected by law, they continue to be hunted for their beautiful coat and for their bones and teeth, which some Asian cultures believe to have healing powers.

Researchers believe that mysterious clouded leopards live solitary lives, other than mothers caring for their cubs.

Red River Hog

Red river hogs are named for their reddish-brown fur and the fact that they often wade through water. These colorful hogs are active both day and night and are surprisingly good swimmers, holding their tail above the water as they paddle. They can also swim underwater, surfacing to breathe every 15 seconds or so.

Researchers believe that the red river hogs' large, tufted ears may be used in territorial or dominance displays. The pigs use their tough snout to root up tubers, roots, insects, and worms. As herd animals, they are social with their keepers and are easy to train, learning tricks relatively quickly.

DID YOU KNOW?

Red river hogs have a flexible snout that is strong enough to dig up large boulders. Like all wild pigs, their tail is straight. Only domesticated pigs have a curly tail.

Thorn Acacia

The San Diego Zoo has long cultivated acacias for beauty, shade, and browse for its hoofed animals. With 38 species, acacias make up one of the Zoo's accredited plant collections.

Acacias have been an important part of human history since ancient times. They're known not just for their shade, waterproof wood, gum arabic, and pleasant scent, but also for their reputed oracular and magical powers. Ishtar, the Mesopotamian goddess of love and war, used acacias for her oracles. Jehovah, the Hebrew god, had his sacred Ark of the Covenant built from acacia wood. And acacias played a significant role in the resurrection of Osiris, the Egyptian god of the underworld.

About 800 species of acacias inhabit the earth's warmer climes, especially Australia, where the trees are called wattles. Because of their usefulness, acacias are cultivated in many areas where they are not native.

CATCH A GLIMPSE

The thorn acacia *Acacia albida* is easy to spot—just look for the tallest tree in Urban Jungle. This thorny tree, native to Africa and the Middle East, is located across from the giraffe exhibit.

Australian Outback

Koala

Tasmanian devils are the top carnivores in Tasmania and will travel up to 10 miles each night in search of prey and carrion.

Australian Outback

The Conrad Prebys Australian Outback features a wide variety of Australian birds, plants, and mammals. Lining the entrance are towering aboriginal-style totem poles, each topped with an iconic Australian species. A path then leads guests through a eucalyptus grove, where species like wallabies, cockatoos, kookaburras, and Tasmanian devils live.

The Queenslander House is across the road, with wide decks that overlook the Zoo's koala colony—the largest breeding group outside of Australia. Guests can see the koalas at eye level from the deck. Because the males can be territorial, they have their own perch with dividers separating them. The House itself functions as an education classroom and has displays about koala conservation, a window into the koala food prep kitchen, and information about how the Zoo cares for the koalas.

Star finches spend most of their time in small groups, foraging for seeds in Australia's grasslands and marshlands.

Koala

Koalas might look soft and cuddly, but they are wild animals! Their fur actually feels like the wool on a sheep. Koalas have two thumbs on each hand and ridged skin on the bottoms of their feet that give them traction for climbing. With strong arm and shoulder muscles, they can climb 150 feet to the top of a tree and leap easily from treetop to treetop.

As marsupials, female koalas carry their babies—called joeys—in a pouch. Newborns are about the size of a large jellybean and unable to see or hear, but they can crawl from the birth canal into their mother's pouch. There, they attach to a nipple and nurse. After about six months, joeys venture out, returning to the pouch when they want to hide or sleep. When they're too large for the pouch, they climb onto their mother's back but may still stick their head into the pouch to nurse. They become fully independent around one year old.

Koalas only eat eucalyptus leaves, consuming about one pound of leaves each day. Eucalyptus leaves are poisonous to most animals, but koalas have bacteria in their stomach that breaks down the toxic oils. They also have special cheek teeth that help them grind the tough leaves. Since koalas don't get many calories from their diet, they conserve energy by moving slowly and sleeping up to 20 hours a day. They live in the trees with their rear end firmly planted in the fork of branches so they can eat and nap without feeling threatened.

Behind the Scenes

Through the San Diego Zoo's Koala Education & Conservation Program, koalas are sent on loan to other approved zoos so that more people can see and learn about these charismatic marsupials. The koalas that go on these loans are selected because of their calm and flexible personality, so the new surroundings are not a problem for them. A keeper travels with each koala and stays with the animal until it is settled in.

Koalas are complex and sometimes unpredictable creatures. The Zoo's koalas each have a distinct personality, and some are named for a particular characteristic or trait.

Wallaby

Wallabies are smaller, stockier members of the kangaroo family. These herbivores eat grasses, vegetables, leaves, and herbs. Like other kangaroos, wallabies use their strong tail for balance and support—and their powerful hind legs to bound at high speeds and fend off potential predators with a vigorous kick.

The San Diego Zoo has parma wallabies on exhibit. Native to the forests of Australia, parma wallabies are nocturnal and usually shelter in thick scrub during the day. They create runways through the brush and can travel quickly through them if needed. Although they're mostly solitary, a few may come together for feeding or around a watering hole in the dry season. Parma wallabies are shy creatures. At the Zoo, the females might warm up to their keepers, but the males will remain aloof.

Parma wallabies were considered extinct until 1965, when workers on Kawau Island, New Zealand, discovered that what they thought were tammar wallabies overrunning the island were actually a surviving population of parma wallabies. The wallabies were caught and sent to managed breeding facilities around the world in the hope that they would reproduce and eventually be reintroduced to their native habitat. Two years later, another population was found in the forests of Gosford, Australia. Today, while parma wallabies are still rare, their populations are considered stable.

FIELD NOTES

Like other marsupials, wallabies mate throughout the year, but females can keep their embryo in suspended animation until resources are plentiful or their most recent joey leaves the pouch. Mothers can also manufacture two kinds of milk for their overlapping offspring. Females can simultaneously have a joey "at heel" and nursing, another in the pouch nursing, and an embryo in suspension in her uterus.

Tree Kangaroo

DID YOU KNOW?

Tree kangaroos have wooly fur that protects them from the rain showers in their habitat. They even have a whorl of fur on the back of their neck that redirects water and keeps it from dripping onto their face during naps.

Tree kangaroos are the only arboreal members of the kangaroo family. While typical kangaroos have large, muscular hind legs and smaller forelimbs, tree kangaroos have longer, stronger, more flexible forelegs and a long, flexible tail that helps with balance. They also have long, curved claws and spongy gripping pads on their paws to navigate the trees. Unlike other kangaroos, they can move their back legs independently of one another and move backward.

Tree kangaroos are found in dense rain forests where they travel rapidly from tree to tree, sometimes leaping as far as 30 feet down into the branches of another tree. Most kangaroos and wallabies are grazers, clipping grasses on the ground with their teeth, but tree kangaroos eat directly from the trees, consuming leaves, flowers, ferns, mosses, and occasionally insects. Like other macropods, they have a chambered stomach that enables them to bring the vegetation they've recently swallowed back up from one chamber, chew it as cud, and then swallow it again for final digestion. Their leafy diet requires time to digest, so tree kangaroos sleep quite a lot. They wedge themselves against tree branches, drop their head onto their chest, and snooze away.

Tasmanian Devil

Native to the forest, woodland, and agricultural areas of the island state of Tasmania, these meat-eating marsupials are feisty and frequently misunderstood. Their reputed ferocity—and even their name—comes from some of the sounds they make. It's said that early European settlers proclaimed that the creatures had to be the devil himself when they heard the eerie growls the animals made when searching for food, as well as the screeching and spine-chilling screams they emitted when feeding on a carcass.

Although Tasmanian devils are rather small, measuring 23 to 26 inches long, they are the largest surviving carnivorous marsupials. They hunt small mammals and birds, mostly at night, but are major carrion eaters, scavenging anything that comes their way. Solitary by nature, their feisty nature comes out when a number of them come together at a carcass.

Tasmanian devils are spirited from the start. In fact, their reproduction takes "sibling rivalry" to a whole new level! Females give birth to anywhere from 20 to 30 young in one litter. About the size of a grain of rice at birth, the joeys must "race" a distance of about three inches into their mother's pouch. There, they compete to locate and attach to one of four teats. Only the four that latch on will have a chance to survive.

Mission Conservation

Once widespread through Australia, Tasmanian devils are found only on Tasmania. For years, they were threatened by hunting. Devils gained legal protection in 1941, but now, they face another serious challenge: devil facial tumor disease (DFTD). This rare, contagious cancer is found only in devils and is transmitted through biting, which commonly occurs when the devils mate and feed. There is no known cure or vaccine. The San Diego Zoo is a proud partner of the Save the Tasmanian Devil Program based in Tasmania, which collaborates with research institutes and zoos around the world.

Wollemi Pine

With their fern-type fronds, multiple trunks, and textured bark, Wollemi pines are unusual-looking trees—and one of the rarest plants in the world. With fewer than 100 left in the wild, these coniferous trees from the Blue Mountains are listed as critically endangered. Although they are commonly referred to as "pines," they're not pines at all: they are part of the Araucariaceae family. In their native habitat, these trees can withstand extreme temperatures and reach heights of 125 feet.

Previously known from 120 million-year-old fossil records, Wollemi pines were believed to be extinct until 1994, when a park ranger found a stand of less than 100 trees in a remote location in Wollemi National Park, west of Sydney, Australia. Today, horticulturists and scientists are working to ensure the species' survival through captive propagation.

CATCH A GLIMPSE

Two young Wollemi pines are growing between the bird aviaries across from the Tasmanian devil exhibit. As these trees mature, their bark will take on the "bubbled" texture that is a unique characteristic of these "living fossils."

To protect the remaining wild Wollemi pines, scientists and horticulturists are keeping the trees' exact location in the rugged Blue Mountains a secret.

Cockatoos are mostly black, white, or light pastel pink due to the lack of a special structure in their feathers. In other parrots, this structure produces color by the way it reflects light.

Palm Cockatoo

FIELD NOTES

Cockatoos are noisy! They scream to communicate with one another—as well as simply to make noise. To advertise their territory, palm cockatoos hold a stick in their foot and drum it against a hollow tree to make a reverberating sound. Males also do this to attract females during breeding season.

With their perky crest and natural curiosity, cockatoos are among the most well-known and loved members of the parrot family. Like all parrots, cockatoos can use their feet much like humans use their hands, making them strong climbers—an important adaptation that enables them to access fruit and nuts high in the trees. Cockatoos can even hold a piece of food in one foot while balancing on the other.

Palm cockatoos, also known as goliath cockatoos for their large size, are found in the rain forests and woodlands of northern Queensland, Australia, and New Guinea. With their powerful bill, they can eat hard nuts and seeds that other species have difficulty managing. They primarily feast on the fruit and nuts of the *Pandanus*. Unlike other cockatoos, palm cockatoos are not flock feeders. They are usually observed alone, in pairs, or in groups of up to seven. At sunset, small groups of palm cockatoos will separate from the flock with their partner and return to their own territory to roost.

Eucalyptus

There are about 700 species of eucalyptus. Most of them are native to Australia, where the trees provide food and shelter for many types of animals. There are six main types of eucalyptus, with varying textures of bark. The well-known gum trees—made popular through a children's song—have smooth bark. The other types have rougher coverings.

More than three-quarters of the forests in Australia are comprised of eucalyptus trees, a range that covers approximately 450,000 square miles. Eucalyptus trees were originally introduced to California in 1856, where they became popular because they grew quickly, didn't need a lot of water, and could be used for shade, wood, and products made from the tree's aromatic oil. At the San Diego Zoo, the eucalyptus trees add to the lush landscape and provide food for the koalas. The animals receive 34 different eucalyptus species that are harvested from the Zoo, the Safari Park, and the Zoo's browse farm.

Wombat

DID YOU KNOW?

Giant wombats lived during the Ice Age and were the size of a rhinoceros. It's believed that ancient Aborigines hunted them for food.

Wombats are most closely related to koalas. But while koalas are suited for trees, wombats are more comfortable on the ground. With their wide, strong feet and large claws, wombats are masterful diggers. In fact, unlike most other marsupials, the females have a pouch that faces their rear end to prevent dirt getting into it while they dig.

These mammals often move up to three feet of dirt in a single night, creating impressive tunnels underground that can reach up to 650 feet in length. Their burrow usually has one entrance that branches out into several tunnels that lead to sleeping chambers. Wombats dive headfirst into the entrance to their burrow when being pursued by predators like dingoes or Tasmanian devils—leaving only their tough, leathery rear end for the predator to try to bite.

With their rodent-like teeth and strong jaws, wombats can grip and tear food like grasses, roots, shoots, tubers, and even tree bark. A special stomach gland helps them easily digest their tough food. Wombats don't need much water, getting most of their moisture from the plants they eat. They are often seen grazing at night, when their coloration helps them blend in, but they may also feed during the day if it's cool and cloudy.

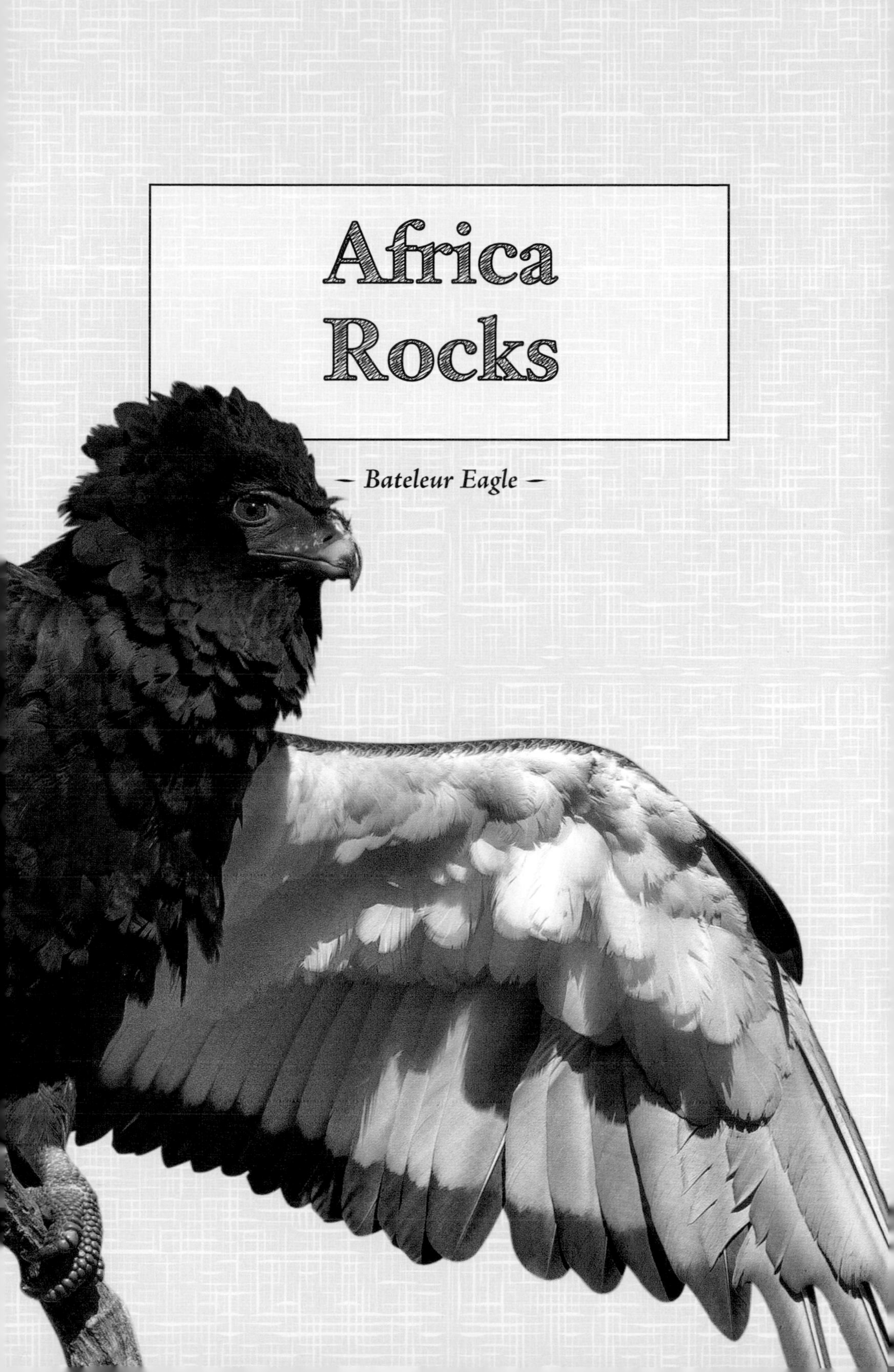

Africa Rocks

– Bateleur Eagle –

Like klipspringers, rock hyraxes (left) are built for life on the rocks. They have a moist, rubbery pad on the bottom of each foot that creates a suctioning effect and helps them cling to smooth surfaces.

Africa Rocks

As one of the Zoo's oldest areas, Africa Rocks is in transition with major plans to showcase a variety of African habitats and wildlife. The area is currently home to various savanna animals, from meerkat colonies to big and small cats. Here, klipspringers bound across boulders on the tips of their hooves, and rock hyraxes scurry in and out of crevices. On the opposite side of the narrow canyon is Big Cat Trail, where lynx and different species of leopards live.

Africa Rocks features a kopje, a rocky island in the middle of an African grassland. Kopjes are complete ecosystems, supporting plants and animals that would be unable to survive outside the protection of the rocks. These "islands" are way stations for weary animal travelers, providing a cool respite from the parched dryness of the surrounding savanna.

It can take bateleur eagles up to eight years to develop their adult black-and-white plumage.

Klipspringer

Klipspringers are famous for their rock-climbing and jumping abilities. Like all antelope, their rounded hooves are split down the middle, dividing the hoof into two "toes." But klipspringers also have a pad in the center that acts like a suction cup, allowing them to hop from rock to rock without falling.

Unlike other antelope species, klipspringers have a short and muscular body that enables them to jump up rock walls that are nearly vertical. This gives them an advantage in escaping their predators, which include lions, leopards, eagles, and even humans. Their speckled coat provides camouflage on the rocky kopje, where they often "freeze" for long periods of time to avoid detection.

FIELD NOTES

Klipspringers are well adapted to life in their rocky homes. These herbivores eat plants that grow in the rocky terrain, and they never need to drink water—they get all the moisture they need from succulents. When lush green plants are scarce during the dry season, they can also digest tough, woody plants.

Sausage Tree

Large, scarlet, and trumpet shaped, the flowers of the sausage tree (*Kigelia pinnata*) are full of nectar that attracts birds, bees, and even baboons. The tree's name comes from its fruit: pendulous cylinders up to two feet long that resemble sausages hanging from a deli ceiling. But despite their tempting name, the pods are not edible. They may be very useful in other ways, though: in Africa they are sliced and applied to heal skin abrasions, and they have been found to contain the chemical lapachol, which may be effective against skin cancer.

Meerkat

Meerkats may look like prairie dogs on a diet, but they are actually related to the mongoose. They often dig in the dirt for insects to eat or to add to their network of underground burrows. These little carnivores live in a group called a gang, or a mob. They work together for the common good, whether helping to care for youngsters or looking out for danger.

Meerkats on lookout duty balance on their hind legs. A meerkat sentry doesn't miss much. If trouble is spotted, the lookout alerts the others, and they all dash back into the safety of the nearest burrow. When they're not on duty, they relax on their backside with their front paws draped across their stomach.

Meerkats on lookout make a low, constant peeping sound when all is well and a bark or whistle when a predator is spotted.

BEHIND THE SCENES

It can be challenging to build a habitat that allows the Zoo's meerkats to dig to their heart's content without digging their way out. The bottom of the Zoo's two meerkat enclosures are lined with wire mesh. (The other enclosure is in Discovery Outpost.) Each exhibit also has large rocks or a man-made termite mound for the meerkat guard to perch atop, vigilantly looking for birds of prey.

Rock Hyrax

DID YOU KNOW?

With their excellent vision, rock hyraxes can spot a predator more than 1,000 yards away.

While rock hyraxes resemble large guinea pigs or rabbits, they are neither. These small animals share an ancestor with the elephant! Like elephants, hyraxes have strong molars that grind up tough vegetation, as well as two large incisor teeth that extend just past their lips like tiny tusks.

Hyraxes hang out in areas that have boulders, rock formations, or even little nooks on sheer cliffs that provide shelter and protection. They often feed in a circle with their head pointing outward to look for predators. While feeding, the dominant male pauses between bites to watch for danger, sounding a shriek alarm if he sees anything of concern. This sends all the hyraxes scrambling for cover, where they remain absolutely still until the danger has passed.

Within each colony, there is one male for every five to seven females. Males leave their birth troop to find and head up a troop of their own; females usually stay in their birth group for life. Hyraxes weigh just a few ounces at birth but look like miniature adults. Within three days, they begin eating some solid foods along with their mother's milk. Hyraxes have special bacteria in their stomach that helps digest tough plant material. Baby hyraxes get "dosed" with the bacteria by eating the adults' droppings.

Rock hyraxes start each morning with a long sunbathing session. They are active for just 5 percent of the day.

Bateleur Eagle

With their black feathers, snow-white underwings, and bright red face and legs, these birds are a sight to behold. The name bateleur is French for "acrobat or tightrope walker," referring to their daring aerial maneuvers during courtship and wing-tipping habits in flight. During courtship, males somersault 360 degrees and females fly upside down. These eagles mate for life, spending up to nine hours each day seeking out mice, birds, antelope, snakes, and carrion on the savanna below. In spite of their hunting prowess, only about 2 percent of bateleur eagle chicks survive to adulthood.

FIELD NOTES

Bateleur eagles build their nest of sticks and twigs, usually in the fork of a tall tree. They have even been known to nest on top of another bird's structure! The wing and tail feathers of juvenile bateleurs are longer than an adult's. The extra length is believed to give the youngsters more stability as they learn to fly. As they molt each year, their new feathers grow in shorter.

Snow Leopard

Shy and secretive, snow leopards live in mountains as high as 18,000 feet. They are almost impossible to study in the wild because they blend in so well with their surroundings and live in such extreme conditions. They are well adapted to their environment, with huge, furry paws that provide traction and cushion their feet; a long tail for balance; and smoky gray and blurred black markings that provide superb camouflage in the rocky mountains.

These stealth predators are capable of killing animals up to three times their weight. Problems arise in the winter when marmots—their primary food source—hibernate. This causes snow leopards to turn to livestock for food, creating conflict with herders and farmers. The more that marmots are hunted for their pelts and meat or killed as pests, the more that snow leopards must hunt livestock in order to survive.

Mission Conservation

It is estimated that there are no more than 4,000 to 6,500 snow leopards over their entire range. Although they are endangered, they continue to be hunted—illegally in many areas—for their beautiful fur and their bones, which are used in traditional Asian medicine. Cooperation between governments, conservation agencies, and the general public is essential. Fortunately, there are efforts in place to protect snow leopards. San Diego Zoo Global participates in the Species Survival Plan for the snow leopard and provides direct support to the Snow Leopard Trust. More than 100 protected areas now exist for these cats, 36 of which are found within international borders.

Snow leopards can pounce on prey as far as 45 feet away. In the wild, they bring down prey every 10 to 15 days.

When a lynx cub first opens its eyes at about two weeks old, its irises are bright blue. The color changes to greenish-gold after many months.

Siberian Lynx

Lynx are known by the tuft of black hair on the tips of their ears and their short, or "bobbed," tail. Some researchers think lynx use this black hair like whiskers to feel things around them. They also have a mane of longer hair around their face and neck.

Lynx are solitary, with males and females only coming together for breeding purposes. Young lynx usually live with their mother for about a year. Sometimes siblings that have just left their mother's side will travel and hunt together for several months before going their separate ways. They reach adult size when they are two years old.

Lynx regularly hunt prey three to four times their size and can even kill reindeer when given the opportunity. The cats rarely chase after potential food, especially if the snow is deep. Instead, they hide behind tree stumps or rocks until a potential meal walks by.

CATCH A GLIMPSE

Look for the Zoo's lynx lounging under the large honey-suckle close to the front of their exhibit. These cats are naturally camouflaged to hide in plain sight, so take the time to look carefully—they are probably closer than you think!

Lynx have fur on the undersides of their feet. This gives them traction on snow and makes their footsteps more quiet so they can stalk their prey.

Amur Leopard

FIELD NOTES

Although they're powerful and clever hunters, leopards are not always at the top of the food chain. Their predators include lions, hyenas, and wild dogs in Africa, and tigers in Asia. Leopards go to great lengths to avoid these predators, hunting at different times, pursuing different prey, and resting in trees to keep from being noticed.

Leopards are the smallest of the large cats. These nocturnal cats look similar to jaguars, with flower-shaped spots on their back called rosettes, but unlike jaguars, they don't have dots in the center of their rosettes. The spots help them hide from their prey, breaking up their body outline in forests or grasslands.

Leopards are the most arboreal of the large cats, with a long tail that helps them balance on narrow tree branches. They're also incredibly strong, able to climb as high as 50 feet up a tree while holding a fresh kill in their mouth—even one larger and heavier than themselves. They stash food high in trees so other predators such as tigers can't get it. This way, they can return to eat more.

Leopards use their vision, keen hearing, and whiskers while hunting. Their whiskers face forward while walking and are pulled back when sniffing. They stick out sideways when resting.

Of the eight leopard subspecies, Amur leopards are the most critically endangered with only about 40 remaining in the wild and approximately 300 in zoos and other managed settings. They are threatened by poaching, loss of habitat, and inbreeding due to tiny, isolated populations. The Amur Leopard and Tiger Alliance (ALTA) and its partner, Wildlife Conservation Society (WCS), have been monitoring the leopards since 1997, using tracking data to estimate their population size. In 2002, they began adding camera traps to identify individual leopards and monitor them over time. Fortunately, antipoaching efforts and educational programs appear to be working, and the population—though small—has remained stable.

Mission Conservation

The San Diego Zoo participates in the Species Survival Plan for Amur leopards, carefully monitoring its three Amur leopards as a vital part of the captive population. When the time is right, the Zoo's cats will be paired with unrelated leopards and relocated to other national or international facilities. As a conservation organization, the Zoo is working with other facilities to develop a sustainable and genetically diverse population of Amur leopards that can contribute to new scientific knowledge and the survival of the species.

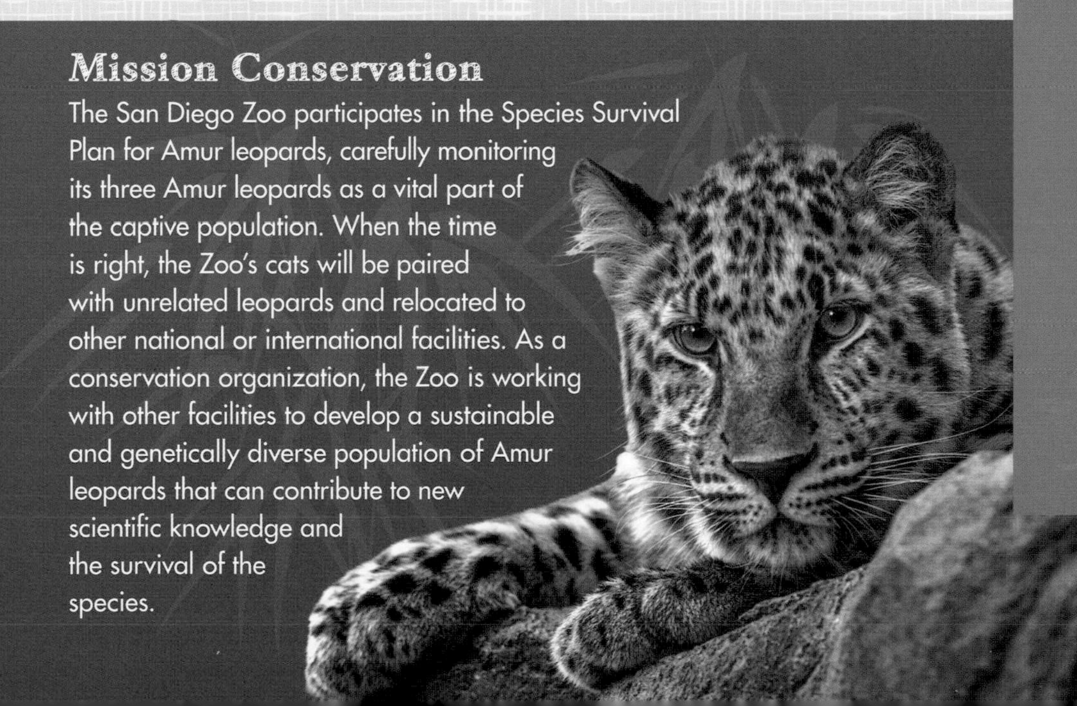

Also known as the shining masdevallia, the orchid species *Masdevallia norops* is native to Ecuador.

Orchid House

The largest family of flowering plants, orchids are found on every continent except Antarctica and in almost every kind of habitat. Some orchids grow in rotting wood, and others bloom underground. The orchid family has more than 35,000 known species and at least another 100,000 hybrids. At the Zoo, there are 948 taxa; 22 of the orchid species are conservation worthy.

The Zoo's Orchid House includes more than 3,000 orchid plants from over 900 species, plus many varieties and cultivars. It has both cold rooms and hot houses, since orchids thrive in many different kinds of temperatures. The Orchid House is open to Zoo visitors during Orchid Odyssey and Plant Day on select days (usually the third Friday of each month). During this time, the Zoo's orchid expert and members of the San Diego County Orchid Society are available to answer questions. The Orchid House is located on the path just off Front Street in the Africa Rocks zone, near Map Locator 10.

CATCH A GLIMPSE

In addition to the Orchid House, one of the best places to view the Zoo's orchids is in the treetops of Fern Canyon and the display case at the bottom. They also grow in the orchid garden at Albert's Restaurant.

The *Phalaenopsis* hybrid, or moth orchid, is the most common orchid and can be easily grown at home.

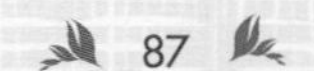

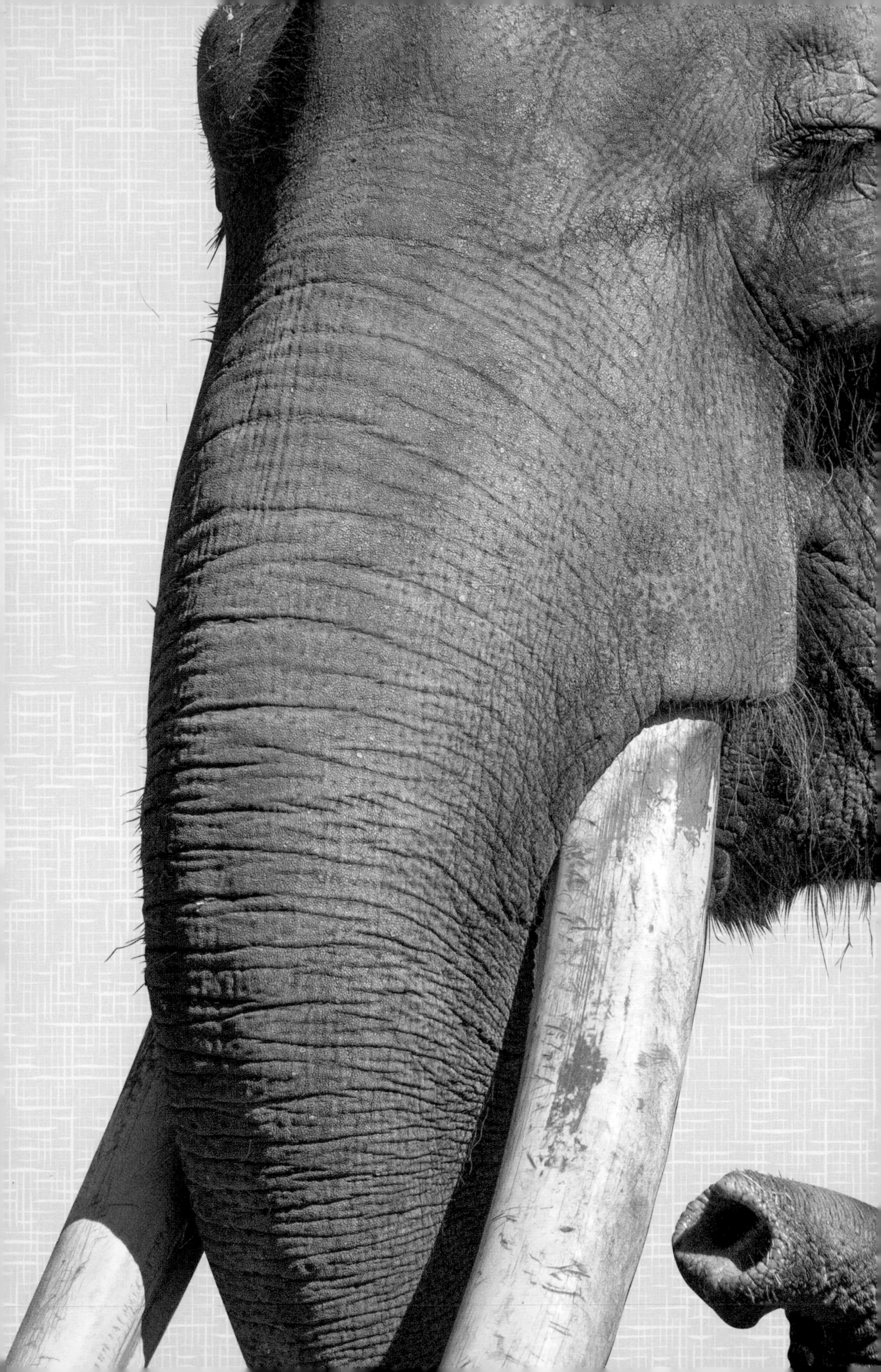

Elephant Odyssey

– *Elephant* –

The Zoo's elephants love the exhibit's swimming pools and mud wallows. On warm days, they can sometimes be seen splashing and spraying each other with trunks full of water.

Elephant Odyssey

The Harry and Grace Steele Elephant Odyssey is one of the largest exhibits at the San Diego Zoo. Spanning from the Pleistocene epoch 12,000 years ago to the present, it reveals Southern California's ancient past through animals that exist today. The mammals, birds, reptiles, amphibians, and insects at Elephant Odyssey may seem unrelated, but they're connected through their ancestral history in Southern California. Some barely resemble their prehistoric counterparts, while others are easy to identify as relatives. Still others died out in North America but flourished elsewhere.

At Elephant Odyssey, visitors can examine fossils, learn about natural history, and see a variety of animals, from elephants and lions to dung beetles and capybaras. Zoo guests can also learn what is being done to preserve these species for the future.

The skin on a condor's neck turns deep red-pink during courtship or when the bird is excited or alarmed.

Elephant

FIELD NOTES

Calves "inherit" their mother's status within the herd. Bottom-ranked animals are submissive to their elders and the other calves that outrank them. Instead of walking head-on into a dominant group of elephants, lower-ranking elephants turn around and back into the space. They also get out of the way when a more dominant elephant is walking around. Elephants must obey the social rules or risk upsetting their 6,500-pound superiors!

Elephants are found on the continents of Africa and Asia. At first glance, African and Asian elephants look the same, but they're actually two different species. African elephants have very wrinkled skin and large ears that are shaped like the continent of Africa, while Asian elephants have smoother skin and smaller ears.

All elephants live in herds, greeting each other by putting their trunk in one another's mouth in an "elephant's handshake." They also drape their trunk over each other's back as a sign of affection. A herd is made up of related females, called cows, and their offspring. The matriarch decides when and where the herd will eat, rest, and travel. Adult males, called bulls, don't live in herds. Once males become teenagers, they leave the herd. After they become adults, the bulls visit other herds for short periods of time to breed.

To stay clean and cool, elephants spray themselves with water. To drink, they draw water into their trunk and squirt it into their mouth.

Newborns weigh a hefty 200 to 270 pounds at birth. Calves stick close to their mother and nurse frequently. The whole herd nurtures the calves, forming protective circles around the youngsters if the adults are startled by new noises or scents.

Elephants have incredible features: their trunk has about 40,000 muscles—more than a person has in his or her whole body. In fact, their trunk is so strong and agile that it can push down trees or pick up a single piece of straw. Their tusks are used for defense, digging for water, and lifting things. And their ears are a little like air conditioners. The skin on their ears is thin, so when they flap their ears on a hot day, they release excess heat from the blood that flows through their veins.

Mission Conservation

Elephants are threatened by poaching due to the sale of ivory on the black market. There are only about 450,000 African elephants and fewer than 50,000 Asian elephants alive today. The San Diego Zoo has teamed with Elephants Without Borders (EWB) to study and save over 220,000 elephants that live in southern Africa. Leading the effort is Dr. Mike Chase, one of the world's top elephant ecologists, founder of EWB and a Fellow at the San Diego Zoo Institute for Conservation Research. EWB has deployed GPS collars on dozens of elephants to track their movements. To learn more, visit endextinction.org.

California Condor

Many Native American tribes have great respect for California condors and see them as symbols of power. In legends, condors were called "thunderbirds" because they were thought to bring thunder to the skies with the beating of their huge wings.

When they fly, California condors are a wonderful sight to behold. Their impressive wings catch thermal air currents that rise up as the sun heats the ground. They can stay aloft for hours as they scan the ground below, looking for food. Like all vultures, California condors feed on dead animals such as deer, cattle, and sheep as well as rodents, rabbits, and fish. In coastal areas, they also feed on the carcasses of seals, dolphins, and whales. As scavengers, they are part of nature's cleanup crew and an important part of the ecosystem.

Some people think of vultures as "dirty," but California condors are actually fairly tidy. After eating, they clean their head and neck by rubbing them on grass, rocks, or branches. Condors also bathe frequently and spend hours smoothing and drying their feathers.

DID YOU KNOW?

California condors are gray-headed as young birds. Their head turns a distinctive orange at four to six years of age. Their bald head and neck prevents food from sticking to them while they eat.

Secretary Bird

These African birds of prey look like cranes but act more like eagles. In fact, secretary birds are so unique—both in looks and how they capture prey—that they're in their own scientific family.

Secretary birds are the tallest birds of prey and the most terrestrial. They stalk reptiles, insects, and small mammals on the ground, hunting on foot through tall grass in search of small mammals, birds, snakes (even venomous ones), and large insects.

There are two theories as to why they are called "secretary birds." Some people say their long tail feathers and head plumes reminded early biologists of a 19th-century male secretary wearing a gray tailcoat with a quill pen tucked behind his ear. The more likely theory is that the word *secretary* derives from an Arabic term meaning "hunterbird."

Secretary birds can cover up to 20 miles each day on foot searching for food. At night, they often roost high in acacia trees.

Jaguar

Jaguars are members of the "big cat" group, along with lions, tigers, and leopards. Because they're both spotted, jaguars and leopards look a lot alike to some people. But here's the secret to telling them apart: jaguars are stockier and heavier with a shorter, thicker tail. Their dark spots, called rosettes, have irregular, broken borders and often a spot in the center. While leopards also have dark rosettes, they lack the center spot, and the edges of their rosettes are unbroken. Most jaguars have tawny-colored fur with black rosettes, but some are black-on-black. Usually jaguars that are found in darker rain forest areas are black.

Like other cats, jaguars have a mirror-like structure in the back of each eye that reflects light into the retina, nearly doubling their ability to see. Jaguars see less detail and color in daylight but have better vision at night.

Jaguars stalk and ambush their ground-dwelling prey at night instead of chasing it like cheetahs and lions do. Their large jaw muscles allow them to pierce their prey's skull with their sharp teeth. This allows them to eat spectacled caimans and hard-shelled reptiles like turtles and tortoises. Researchers have counted over 85 species in the jaguar diet, including peccaries, deer, tapirs, cattle, and capybaras.

Mission Conservation

There are an estimated 10,000 jaguars left in the wild, and their numbers are decreasing rapidly. Poachers hunt jaguars for their coat, while loggers, miners, and farmers devour their habitat. As wild prey becomes less available, jaguars sometimes kill livestock for food and are killed in response. The San Diego Zoo is developing programs to help ranchers avoid problems with the cats. The Zoo is also working with Latin American scientists to study, monitor, and protect jaguars. Researchers are using camera traps—which take a photo when the animal crosses in front of the camera—and are placing radio collars on some jaguars to track their daily movements.

The South American native word for jaguar, *yaguará*, means "animal that kills in a single bound."

Capybara

Capybaras are the largest rodents in the world. They share some features with mice, squirrels, and porcupines—especially their long, sharp front teeth. Capybaras graze on grass and water plants, eating six to eight pounds of grass every day. Like goats, cows, and giraffes, capybaras will regurgitate their food in order to chew it some more. And they tend to chew in a side-to-side motion, like camels.

Called "capys" for short, these endearing animals are found on Central and South American riverbanks or wherever standing water is available. Besides finding food in this habitat, capys use the water itself to escape from danger. Like hippos, their eyes, ears, and nostrils are all located near the top of their head, which helps them learn everything they need to know about their surroundings while the majority of their body remains hidden underwater. They can stay underwater for up to five minutes to hide from predators.

Capybaras are social animals, typically living in groups of 10 to 30 animals. Dominance is important for group living, but capys are usually peaceful. They communicate using barks, chirps, whistles, huffs, and purrs, chattering back and forth to keep track of one another. Capybaras also communicate via a pair of scent glands on their rump that they use to mark their territory. The males even mark females! Females scent mark less often than males do, mostly during the wet season when they are looking for a mate.

CATCH A GLIMPSE

Look for the male with the biggest *morrillo,* or scent gland, on the top of his nose. The bigger the morrillo, the more dominant the male. The most dominant male is usually found in the center of the group, with access to the best food and the females.

Juvenile capybaras—which are often preyed upon by ocelots and harpy eagles—are nursed and cared for by all the females in their group.

Guanaco

Pronounced "gwa-NAH-ko," these animals are found throughout South America in dry, open country. It's hard not to notice their resemblance to llamas, because llamas—which are strictly domesticated animals—are descended from wild guanacos. Guanacos have large eyes with thick lashes that protect them from dust and dirt. Like their camel relatives, they have two padded toes that help with footing on rocky trails or gravel slopes. Their upper lip is split in two and can be used like fingers to help draw in food. Like cows, these grazers and browsers are ruminants, with a three-chamber stomach. They get all the moisture they need from their food.

Guanacos have some interesting methods of making a statement. They can spit their stomach contents up to six feet away—and they have great aim! They also use their dung to mark their territory.

Mountain lions are their main predator, and guanaco calves are especially at risk. Usually, females in a herd give birth within days of each other. With so many calves on the ground, the babies have a greater chance of not getting taken by a mountain lion. However, only 30 percent of guanaco babies born in the wild will live long enough to become adults.

BEHIND THE SCENES

Guanacos communicate through vocalizations, from high-pitched trilling to snorting and shrieking. The guanacos at the San Diego Zoo sound an alert call when the trainers walk one of the Zoo's animal ambassadors, an arctic wolf, by their enclosure. Even though he's on a leash, he's still a wolf!

Pronghorn

Pronghorns look like antelope, but they're different enough to be classified in their own family. Native to western North America, these critically endangered animals have unique headgear that has qualities of both horns and antlers. The bony cores on their skull are covered by keratin, forming actual horns, like cattle. But true horns are permanent—and pronghorns shed their horn coverings each year, like deer that drop their antlers.

Danger makes pronghorns' hair literally stand on end. Muscle contractions cause their long, white rump hair to flare, flashing a warning that can be seen for more than a mile. Known as "ghosts of the desert," their coloration allows them to blend in with the terrain and vanish at high speeds into the distance, their cotton-white rump patch serving as a beacon so the herd can stick together.

Mission Conservation

San Diego Zoo Global is helping this critically endangered species through a binational effort in Baja California, Mexico. The program provides funding and staff for breeding, monitoring, and management. Translocation sites in Mexico focus on caring for the animals through physical exams, ear tags, microchips, and vaccinations before they are released into the wild. The program has doubled the population in the wild. The Zoo has also worked to hand-raise newborns to place with an "assurance herd" at the Los Angeles Zoo, which helps safeguard against natural disasters and disease.

Native Rattlesnakes

Known for their diamond-shaped head, rattlesnakes are considered the most recently evolved snakes in the world. Their rattle is made of interlocking rings, or segments of keratin—the same material as our fingernails. When vibrated, their rattle creates a hissing sound that warns off potential predators.

Even in total darkness, rattlesnakes can detect their prey's heat. This is due to the heat-sensitive pits on each side of their head. They can strike in darkness if their prey is even slightly warmer than its surroundings. Rattlesnakes only eat about every two weeks, depending on how large their last meal was. Younger rattlers eat approximately once a week.

The word *rattlesnake* fills most people with fear and anxiety. But rattlesnakes help control rodent populations, and their venom has been the focus of many scientific studies worldwide. In fact, rattlesnake venom has many significant pharmacological uses. Rather than be scared, it's important for humans to learn how to coexist with rattlesnakes.

Lion

FIELD NOTES

Prides have a close bond and are not likely to accept strangers. Unrelated males may stay from a few months to a few years, but the older lionesses will stay together for life. Lion researchers have noticed that some activities are "contagious" in prides. One lion will yawn, or groom itself, or roar, setting off a wave of yawning, grooming, or roaring.

An African lion's life is all about sleeping, napping, and resting. In a 24-hour period, lions will have short bursts of intense activity, followed by long bouts of lying around that can total up to 21 hours.

Lions live in prides consisting of several unrelated males, many related females, and their offspring. Hunting as a group means there is a better chance that lions will get food when they need it and makes it less likely that they will get injured while hunting. Lions and lionesses play different roles in the life of the pride. The lionesses work together to hunt and rear the cubs. Being smaller and lighter than males, lionesses are more agile and faster. During hunting, smaller females chase the prey toward the center. The larger and heavier lionesses ambush or capture the prey.

Lionesses are versatile and can switch hunting jobs depending on which females are hunting that day and what kind of prey is involved. It may look like the lionesses do all the work, but the males play an important role. While they do eat more than the lionesses and bring in far less food, males patrol, mark, and guard the pride's territory. Males also guard the cubs while the lionesses are hunting and make sure the cubs get enough food.

A lioness will raise her male cub until he is about two. A female cub, however, may stay with her mother for life.

A lion's mane functions to make him look more impressive to females and more intimidating to rival males. In the wild, it protects him during fights with other males.

Camels make many noises, including moaning and groaning sounds, high-pitched bleats, and loud bellows and roars.

DID YOU KNOW?

Camels can go a week or more without water and several months without food. They can survive a 40 percent body weight loss and then take in up to 32 gallons of water at one drinking session.

Camel

Camels were domesticated more than 3,000 years ago, and humans still depend on them for transport across arid environments. Camels can easily carry an extra 200 pounds while walking 20 miles a day, traveling as fast as horses.

Elephant Odyssey has dromedary, or Arabian, camels. These camels exist today only as domesticated animals and make up about 90 percent of the world's camels. Critically endangered Bactrian camels are both wild and domesticated. Native to China and Mongolia, Bactrian camels can survive a wide range of temperatures, from minus 20 degrees to 120 degrees Fahrenheit.

Dromedary camels have one hump, while Bactrian camels have two. Each hump stores fat, which becomes an energy source. The size of their humps can change depending on the amount of food they eat. When food is scarce, they can survive on the fat stored there.

Camels have a thin, nictitating membrane on each eye. Like a clear inner eyelid, it protects the eye from grit flying in sandstorms while letting in enough light to see. Double rows of extra-long eyelashes help keep sand out of their eyes. And their nostrils can be closed off to keep sand out.

The two parts of their split upper lip can be moved independently, which helps camels pluck short grass. Camels also eat thorns, salty plants, and even fish. Many people know that camels spit, but most people probably don't realize why: spitting is meant to surprise, distract, or bother the animal (or person) they consider to be a threat!

Dung Beetle

Dung beetles consist of about 28,000 species worldwide. They dwell on every continent except Antarctica. They were venerated, embalmed, sculpted, and depicted in hieroglyphs by the ancient Egyptians. In fact, they've been resurrecting pastureland and cleaning up after the animal kingdom for more than 40 million years.

Dung beetles have distinctive clubbed antennae composed of plates that can be compressed into a ball or fanned out to detect odors. They identify their desired food by scent from the air and have been known to fly 10 miles to the perfect pat.

These beetles play an important role in the rapid recycling of organic matter and the disposal of disease-breeding waste. The average cow drops 10 to 12 dung pats per day, and a single manure pat can generate 60 to 80 adult flies. Fly populations have been shown to decrease significantly in areas with dung beetle activity. By burying and consuming dung, these beetles improve nutrients and soil structure, leading to healthier habitats for plants and animals.

DID YOU KNOW?

Dung beetles can move dung balls weighing up to 50 times the animal's own weight.

Elephant-Foot Tree

Certain unique plants and trees from desert or seasonal rainy regions have adapted ways of storing much-needed water in their trunk. These plants are called caudiciforms. Probably the most famous of all caudiciforms is the elephant-foot tree *Nolina recurvata,* formally known as *Beaucarnea recurvata.* The elephant-foot tree is aptly named: the swollen base of its trunk, which holds water, resembles the foot of a pachyderm! Although its frond-like leaves make it look a lot like a palm, it is not. Related to yuccas and agaves, the plant originates in southeastern Mexico and is grown all over the world as a houseplant. In climates like San Diego's, it can be grown outside and become very large.

On most caudiciforms, the foliage is usually small and brittle. This occurs because, in times of drought, many caudiciforms will become dormant, drop their leaves, and live off stored water reserves. *Nolinas,* however, do not drop their leaves but store water in their swollen base. The plant's goal is to make it through the dry period, so little energy is spent on producing vegetative growth. After a *Nolina* flowers, new growth stems bud out and grow.

CATCH A GLIMPSE

The largest elephant-foot tree in this area measures 10 feet wide at the base. For decades, it grew beside the Flamingo Pond at the Zoo's entrance before being carefully moved to Elephant Odyssey. Look for this impressive specimen to the right of the entrance of Elephant Odyssey's tar-pit portal.

Northern Frontier

— Polar Bear —

Gerenuks are well adapted to the African savanna and desert. They can go their whole life without drinking water.

Northern Frontier

PERCHED ON THE EDGE OF THE ZOO'S NORTHERNMOST mesa, the Northern Frontier zone is home to the Conrad Prebys Polar Bear Plunge, where bears swim up to the glass to check out guests and kids can crawl into a polar bear snow den, see how much meat a polar bear eats, or compare their size to a polar bear's. Northern Frontier also boasts a canopy of fragrant conifers that shade agile foxes searching for hidden treats. There's a pond, too, where arctic ducks splash, swim, and sometimes dive. In addition to arctic animals, guests can get a good look at antelope and endangered zebras from equatorial regions of Africa and mountain lions from North America.

Arctic foxes are opportunistic hunters. They will eat anything, including berries, birds and eggs, fish, insects, and even small seal pups.

Polar Bear

Native to the frigid arctic circle, polar bears have developed some amazing adaptations to the cold. In fact, these bears can stay so warm in their habitat that they sometimes overheat and have to cool off in the chilly water! Polar bears have a dense undercoat of fur that is protected by an outercoat of long guard hairs. These guard hairs stick together when wet, forming a waterproof barrier that keeps them dry. Their fat also helps them survive, generating heat to insulate them from the freezing air and cold water.

Polar bears are mainly meat eaters, and their favorite food is seal. As patient hunters, they will stay motionless on the ice for hours, waiting for a seal to pop up from its breathing hole. Their nose is so powerful that they can smell a seal on the ice 20 miles away, sniff out a seal's den that has been covered with

FIELD NOTES

Polar bears are highly intelligent and playful. In the wild, two or more bears sometimes form "friendships" that last for weeks or even years. These bears wrestle as a form of play and may feed and travel together. Individual polar bears have also been observed sliding repeatedly downhill or across ice for no apparent reason other than just for the fun of it.

Even though polar bears look white, their hair is really made of clear, hollow tubes filled with air.

snow, and even find a seal's air hole in the ice up to one mile away. They see well underwater, spotting potential meals from 15 feet away.

Polar bears don't hibernate, but their bodily functions do slow down in the winter. Many scientists call this "winter sleep," because the bears can easily be awakened. Mothers can give birth and nurse their young during their winter sleep.

Newborn cubs are about the size of a rat and are born hairless and blind. They depend on milk from their mother, which is 35 percent fat, the richest milk of any bear species. This helps the cubs grow quickly. Within a few months, they weigh more than 20 pounds and can start exploring with their mother outside the den. At about two years of age, they are ready to be on their own.

Mission Conservation

Polar bears are not only affected by global warming but also industrial human activity. The San Diego Zoo Institute for Conservation Research has studied the effects of noise pollution on the bears' ability to communicate and perceive important cues in their environment. Two polar bears at the Zoo were involved in the study, which revealed their hearing threshold and the sound frequencies they find most sensitive. Researchers hope the results of these studies will create conservation laws to help protect the bears.

Reindeer

Scientists believe that reindeer were first domesticated between 3,000 and 7,000 years ago in northern Eurasia. Still the only deer species to be widely domesticated, reindeer are used as beasts of burden and farmed for milk, meat, and their hide. In comparison to body size, they have the largest and heaviest antlers of all living deer. They are the only deer species in which both males and females grow long antlers.

Reindeer have some interesting adaptations that help them thrive in the cold. For instance, they're covered in hair from their nose to their feet! Having hairy hooves may look funny, but their hair provides a good grip when walking on frozen ground, ice, mud, or snow. They also have long, hollow guard hairs that trap air and hold in body heat. The hollow hairs help them float, too, allowing them to swim when needed.

A social species, reindeer form large regional herds of 50,000 to 500,000 animals during the spring. Herds generally follow food sources, traveling south when food is hard to find in winter. Their superior sense of smell helps them find food hidden under snow, locate danger, and recognize direction. They eat nine to 18 pounds of vegetation a day.

FIELD NOTES

During breeding season, males fight with each other, sometimes to the death. The winner chooses five to 15 females to be in his harem. Those that become pregnant leave the herd in the spring and travel to a traditional calving ground, where they give birth within a 10-day period of each other.

Arctic Fox

DID YOU KNOW?

When food is scarce on the tundra in the wintertime, arctic foxes will often follow polar bears to eat their leftovers.

Everything about arctic foxes helps them survive their cold, harsh habitat. Like reindeer, arctic foxes have hair on the bottoms of their paws to give them good traction as they race across icy ground. Their tail curls around their face like a scarf when the frigid winds blow. Even the color of their fur changes with the seasons of the year. In winter, their fur is white to blend in with the snow. (Although some arctic foxes have a "blue-phase" coat that grows lighter but never turns completely white.) During the spring, they shed their winter coat, revealing gray fur underneath that camouflages them among the tundra's rocks.

Arctic foxes have a great sense of smell and excellent hearing. With their small, pointy ears, they can hear their prey moving around in underground tunnels. When arctic foxes hear their next meal scurrying under the snow, they leap into the air and pounce, breaking through the snow right onto their prey underground.

Grevy's Zebra

Zebras are sturdy, spirited animals that are a study in contrasts: willful and playful, social and standoffish, resilient and vulnerable. Their life in a herd can be complex, yet they also find safety in numbers. They are prey for predators, but they are also able to easily defend themselves.

Endangered Grevy's zebras live in semi-arid grassland in Kenya, Ethiopia, and Somalia. They are the largest of the three zebra species—which include plains and mountain zebras—weighing up to 990 pounds. Interestingly, they have the most delicate striping of all the species, with lines that continue all the way to their hooves. Grevy's zebras recognize one another by their unique striping. They have a thick neck and large, round ears that give them a mule-like physique. Their large ears also help them listen for danger and even allow them to communicate with one another by pointing in the direction of the concern. Unlike other female zebras, which live in harems year round, female Grevy's zebras usually move through the males' territory only during breeding season.

Mission Conservation

About 2,250 critically endangered Grevy's zebras remain in the wild, their numbers decimated by loss of habitat and anthrax bacteria outbreaks. San Diego Zoo Global is a member of the Grevy's Zebra Trust and is working with other groups to preserve the population. The Zoo also supports the Northern Rangelands Trust and the Grevy's Zebra Trust in Kenya, where it has sent staff to help with conservation studies. These local, community-based programs protect habitats for zebras and other wildlife and facilitate population and ecology studies to better understand these species' needs.

While other zebra species are nomadic, Grevy's zebra stallions are highly territorial, marking their range with urine and dung.

Arctic Aviary

Not all arctic animals are big and white; most aren't even furry. The Arctic Aviary at the Zoo includes some interesting and unusual diving ducks and shorebirds, such as buffleheads, smews, and mergansers. With a chilled pond and sandy beaches, the aviary helps these birds feel at home in temperate San Diego. The water in the pond is kept at a comfortable 55 degrees Fahrenheit, which is just right for the birds. There are also barnacles, mussels, and starfish in the icy water, but those creatures aren't real! They need saltwater to survive, and the diving duck pond is filled with fresh water for easier maintenance.

The diving ducks in the Arctic Aviary are highly specialized waterfowl. They overcome their natural buoyancy by reducing the amount of air in their feathers, lungs, and air sacs. This results in less oxygen available for diving, so they slow their heart rate while diving to conserve their scant oxygen supply.

CATCH A GLIMPSE

There are two types of diving ducks in the Arctic Aviary: one bufflehead duck and a pair of long-tailed ducks. Female buffleheads have a white spot behind the eye, and long-tailed ducks have long, thin tail feathers.

Bufflehead ducks usually return to the same nesting site every year: an old woodpecker hole.

Lesser Kudu

Lesser kudu are a species of spiral-horned antelope, typically found in the woodlands of eastern Africa. About half the size of greater kudu, they have a sleek, brownish-gray coat with clearly marked vertical white stripes that make them hard to find in the dappled sunlight coming through trees. Lesser kudu also have white chevrons between their eyes with white patches on their neck. They are not as social as greater kudu and usually live alone or in pairs. They're most active at night.

DID YOU KNOW?

The twisted horns on lesser kudu result from a genetically-directed, controlled growth pulse where the horns grow faster and thinner at certain times and thicker and slower at other times.

Mountain lions generally would rather flee than fight, and they rarely confront humans.

Behind the Scenes

In designing the exhibit for these elusive, powerful predators, the Zoo included plants for privacy and a rock arch where the cats can sun themselves and survey their surroundings. For enrichment, the keepers often provide items the mountain lions can bat and chase, such as dry gourds. The gourds roll unpredictably, and the dried seeds inside make a rattling sound, which entices the cats.

Mountain Lion

DID YOU KNOW?

Despite what you see in the movies, mountain lions don't make that "wild cat scream" very often. They're more likely to whistle, squeak, growl, purr, hiss, and yowl.

Mountain lions, pumas, cougars, panthers—these cats are known by more names than just about any other mammal! In Southern California, they are commonly called mountain lions. As solitary creatures, they live in areas called home ranges, which vary from 30 to 125 square miles. Although these ranges overlap, the cats don't often meet. They mostly leave messages for one another in the form of feces, urine, scratched logs, and marks they scrape out in the dirt or snow.

Mountain lions are powerfully built, with large paws, sharp claws, and muscular hind legs that make them great jumpers. They can run fast and have a flexible spine to help them maneuver around obstacles and change direction quickly. Although they are quick, mountain lions are mostly ambush hunters who stealthily wait to pounce on their prey. They eat a variety of animals, including deer, pigs, capybaras, raccoons, armadillos, hares, and squirrels.

Mountain lions have become scarce across North America. As more people have moved into their territory, the number of encounters with these cats has increased. Some people consider them to be pests and shoot, trap, or poison them. But mountain lions have an essential role to play. They are one of the top predators in their ecosystem, and without them populations of deer and other animals would become unhealthy and too large for the habitat. Problem cats should be reported to local animal control agencies, such as the US Department of Fish and Wildlife.

Gerenuk

The name *gerenuk* (pronounced with a hard *g*) comes from the Somali word *garanug*, which means "giraffe-necked." Like giraffes, gerenuks use their long neck and long legs to reach the best browse (food) overhead. The gerenuks at the Zoo are fed an herbivore pellet diet, but the keepers also hang browse so the animals can feed in their usual way—standing on their hind legs to reach high branches. Native to arid areas in northeastern Africa, reddish-brown gerenuks get most of their water from their food.

Males have long ringed horns that reach 14 to 17 inches, and both sexes have several scent glands. One of these is found in front of each eye—a pit-like opening called a preorbital gland. With their scent glands, gerenuks secrete fluid on grass, twigs, and even other gerenuks to mark their territory.

The gerenuks' range has shrunk over many decades. Although their populations are stable in protected areas, their numbers have continued to decline in locations where they are hunted for their hide.

FIELD NOTES

With their large ears, gerenuks are always listening for the rustling sounds of a predator approaching. They also listen for other gerenuks in their herd. When gerenuks are alarmed, they make a buzzing sound with their mouth to warn the others. They also whistle when they're annoyed and make a loud bleating sound when they are in extreme danger.

Pollinator Garden

DID YOU KNOW?

Thirty percent of the food we eat results from insect pollination—and when you count the insect-pollinated foods that farmers feed their livestock, the number is much higher.

Pollinators such as bees, butterflies, beetles, and flies are in crisis worldwide, suffering from pesticide exposure, habitat loss, and disease. Pollinators make fertilization possible for many plants; without them, food as we know it would not exist. There would be no fruit, veggies, peanut butter, or chocolate—and that's just a start.

The San Diego Zoo is committed to helping pollinators recover. It has created a pollinator way station at the Pollinator Garden located near the entrance to Elephant Odyssey. This space is a safe haven, dedicated to helping sustain pollinators by providing a steady supply of pesticide-free nectar and host plants, as well as suitable living spaces for native bees. For instance, there's a large section of milkweed where monarch butterflies can lay their eggs from spring through fall, helping to boost the West Coast population. This native milkweed is tended by the Zoo's Education Department and Zoo Corps kids.

In addition, the Zoo protects pollinators elsewhere on its grounds by allowing honeybee swarms to move on in their own time. The Zoo only actively removes established hives when either the health of humans or its animal collection is clearly at risk.

Panda Canyon

— Giant Panda —

Pandas learn to climb trees as youngsters and may continue to perch in trees as adults. In the wild, cubs use this technique to avoid predators.

BEHIND THE SCENES

The pandas at the San Diego Zoo receive many types of enrichment. Feeder puzzles made of PVC pipe are stuffed with biscuits and fruit. To get the food items to fall out, the pandas must spin the puzzles around and around, trying to stop them in just the right spot. The keepers also place a variety of scents—including cloves, nutmeg, cinnamon, and perfume—on small blocks of wood and put them into the panda enclosures.

Panda Canyon

PANDA CANYON, WHICH HIGHLIGHTS THE MOUNTAINOUS forests of southwest China, is designed to evoke a researcher's visit to China's Wolong Nature Reserve. It features a replica of a rustic research hut with panda tracking collars, a radio-signal receiver, and reference books, and even "evidence" of the presence of giant pandas, such as green piles of realistic panda scat and paw prints on the ground.

Some of the zoo's most vivid creatures live along these bamboo-lined trails. There are trees and walkways for the russet-coated red pandas, rocky outcroppings for the gold-colored takins, and a shaded enclosure for the rare, green Mang Mountain vipers. At the end of the trail are several viewing areas for the black-and-white giant pandas. These large exhibits include climbing structures, hollowed-out logs, and extra vegetation.

Red pandas are highly intelligent animals. The pair at the Zoo seems to enjoy training.

Giant Panda

FIELD NOTES

Giant pandas have 13 different vocalizations, including a bleat, chirp, honk, bark, moan, growl, squeal, roar, copulation call, and nursing cry. Most of the time, their bleat is used as a greeting during breeding season to exchange information about sex and age.

Giant pandas are a national treasure in China and protected by law. Affectionately called "large bear cats" by the Chinese, giant pandas have been portrayed in Chinese art dating back thousands of years. The rest of the world didn't discover these beloved bears, however, until French missionary Pere Armand David first described them in 1869.

Like other bears, giant pandas are stocky, have a pigeon-toed walk, and are good climbers. But they're unique in many ways: While they have unusually heavy bones for their size, giant pandas are very flexible. (They like to do somersaults!) Even though their diet is 99 percent bamboo, they are technically carnivores. And of course, they have distinctive black-and-white markings that develop shortly after they are born.

Giant pandas inhabit a home range of up to 100 square miles. They spend most of their time eating and sleeping, leaving their territory only when they're searching for new food supplies or a mate. Unfortunately, their natural habitat is shrinking due to deforestation and encroaching development. An estimated 1,600 remain in their habitat, with another 300 in zoos. Worldwide collaborative efforts—including breeding programs, panda reserves, and research initiatives—are in place to help conserve these rare creatures.

Panda cubs stay with their mother until they are about 18 months old. At that point, the mother separates from the cub and will mate again.

DID YOU KNOW?

Giant pandas spend at least 12 hours a day eating bamboo. Because bamboo is so low in nutrients, they need to eat a lot of it—about half their body weight each day!

Although red pandas are largely solitary, they exchange information using scent glands, a variety of calls, and visual cues that include "stare downs" with head bobbing.

Red Panda

Red pandas may share the same habitat, diet, and part of the same name as giant pandas, but that's where the similarities end. While giant pandas belong to the bear family, red pandas are linked to raccoons. In fact, these small, beautiful mammals have been classified as a species all their own: *Ailurus fulgens*, which means "fire-colored cat."

With their reddish coat and white face, red pandas blend in among the red moss and white lichen that cover the tree trunks and branches of their forest home. They spend most of their time in the canopy of old growth trees. With semiretractable claws for climbing and a long ringed tail for balance, they're tree dwellers for a reason. Tree climbing gives red pandas access to the tender tops and young leaves of nearby bamboo stalks, their main source of food. To process bamboo, a low-energy food, they have developed an extraordinarily low metabolism (much like a sloth). This causes them to sleep most of the day, feeding at dawn and dusk.

Red pandas also use their claws for defending their territory. While they're normally mild mannered, they will strike out if threatened. If a swipe of their claws doesn't discourage a predator, they will release a strong odor from the base of their tail.

Mission Conservation

Red pandas are endangered. It is estimated that just 2,500 live in the wild. Their habitat—which consists of old growth forests in China—has been threatened in recent years by logging and farming. Red pandas are also threatened by the pet trade. The San Diego Zoo supports the Red Panda Network, a community-based conservation program in Nepal that hires local forest guardians to maintain camera traps in the forest, talk to local people, and give educational presentations about the importance of preserving the red panda.

Bamboo

Bamboo is one of the world's fastest growing plants, with some species able to grow up to 18 inches a day and reach heights of 100 feet. It is also one of the most versatile plants—used as food for people and animals, timber, paper pulp, musical instruments, toys, fabrics, and medicine. Bamboo may prevent landslides or the collapse of riverbanks where there are erosion problems or during an earthquake. In terms of tension and compression, bamboo is even stronger than wood.

There are more than 1,200 species of bamboo in the wild, but the animals that consume it pick just a few varieties to eat. Both the San Diego Zoo and Safari Park grow about 12 different kinds of bamboo on a combined four and a half acres for the giant pandas, red pandas, and takins. The Zoo harvests more than 10 tons of bamboo each year and stores the grasses in a large walk-in cooler.

While giant pandas bite off large chunks of bamboo stems and leaves in one bite, red pandas daintily nibble one well-chosen leaf at a time. Takins eat bamboo leaves as well as a variety of other vegetation. Eating such a low-calorie food seems impractical, but it can actually be an advantage for these animals, since the plant grows fast and wild on mountainsides and few other species compete for it.

Behind the Scenes

Three times a day, the pandas at the Zoo get a specialized, nutritionist-approved bundle of their favorite food, bamboo. The Zoo provides at least three species of bamboo in every bundle, with a mixture of leafy branches and hard, thick stems called culm. Keepers crack open the thickest pieces of culm to lessen the wear on the pandas' teeth.

Mang Mountain Viper

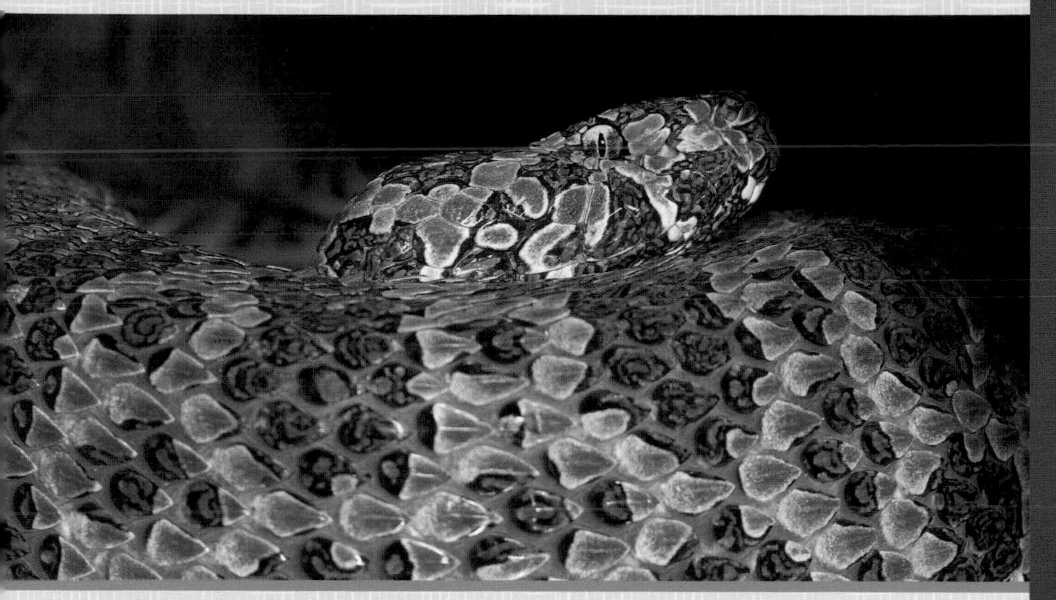

Deadly Mang Mountain pit vipers are nestled in the forests of southern China's Mang Mountain National Forest Park. Reaching lengths of up to seven feet, these snakes have mottled gray-and-green skin that camouflages them against moss-covered logs and rocks while they wait to ambush small prey.

Like other pit vipers, Mang Mountain vipers have a pit between their eyes and nostrils. The opening leads to a special organ that helps them to sense body heat from other animals and judge where to strike. Mang Mountain pit vipers hibernate in deep underground dens during the winter. These snakes were little known outside of China and were not classified by scientists until the early 1990s. The San Diego Zoo is one of only a few places in the world to successfully care for them. In 2012, the Zoo became the first in the country to breed Mang Mountain vipers in captivity, hatching six offspring.

CATCH A GLIMPSE

The Zoo's Mang Mountain vipers can be a challenge to spot because of their excellent camouflage. Take the time to look carefully in crevices, on ledges, and along the bottom of the habitat. It's worth the effort to be able to locate a snake that most people will never see in their lifetime.

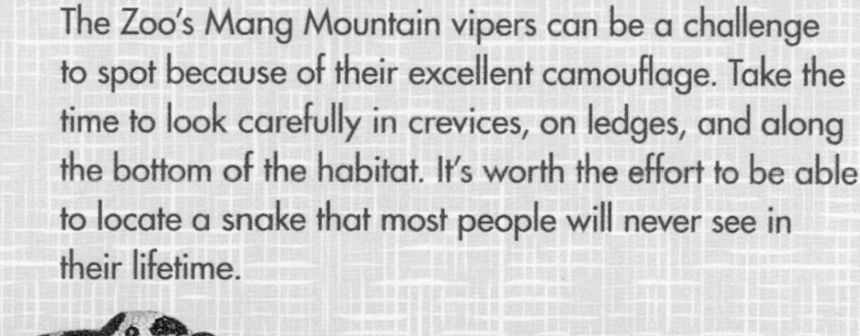

Vipers can control—and conserve—their venom secretion. They can release venom through the left fang, the right fang, both fangs at once, or none at all.

Takin

The Sichuan takin—pronounced "TAH-kin"—has two sharp horns like a wildebeest, a nose like a moose, a tail like a bear, and a large, stocky body like a bison. Adults weigh between 550 and 880 pounds. Despite their hefty size, takins can jump like an antelope six feet in the air from a standing start and leap nimbly from rock to rock.

Takins have some remarkable adaptations to help them survive the bitter cold of winter in the Himalayan Mountains. A thick oil coating on their shaggy fur works like a rain slicker in wet weather, and during the winter, they grow a second, shorter undercoat. Takins also have large nasal passages that warm incoming air and help preserve their body heat.

With their powerful body and impressive horns, these "goat antelopes" have few natural enemies in the wild. Like the giant pandas, however, takins are endangered due to loss of habitat through farming, mining, and logging. They are also vulnerable to climate change. The government of China has given takins full protection under the law and has set aside two reserves for their protection. In 1989, the San Diego Zoo became the first zoo in the Western Hemisphere to successfully breed takins outside of China. Since then, there have been more than 50 Sichuan takins born at the Zoo.

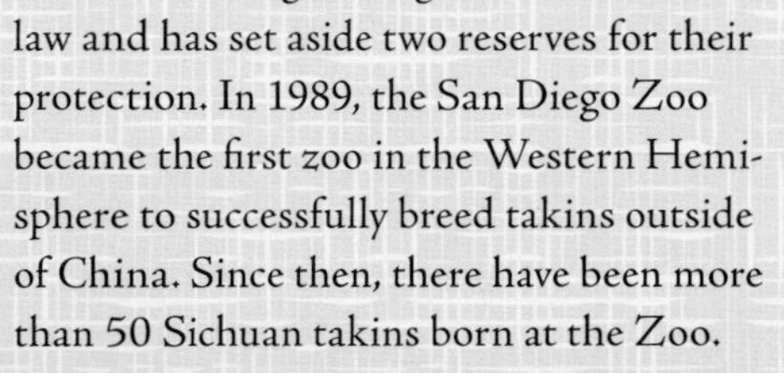

FIELD NOTES

Takins use the same routes over and over in the wild, creating well-worn paths through the dense thickets of bamboo and rhododendrons. Within a couple of days of birth, a takin kid can follow its mother through most types of terrain. If a kid somehow becomes separated from its mother during the first few months, it gives a panicked noise that sounds like a lion cub.

The first Sichuan takin birth outside China occurred at the San Diego Zoo in 1989. Most takins now living at other zoos in North America came from San Diego.

Asian Passage

– Sun Bear –

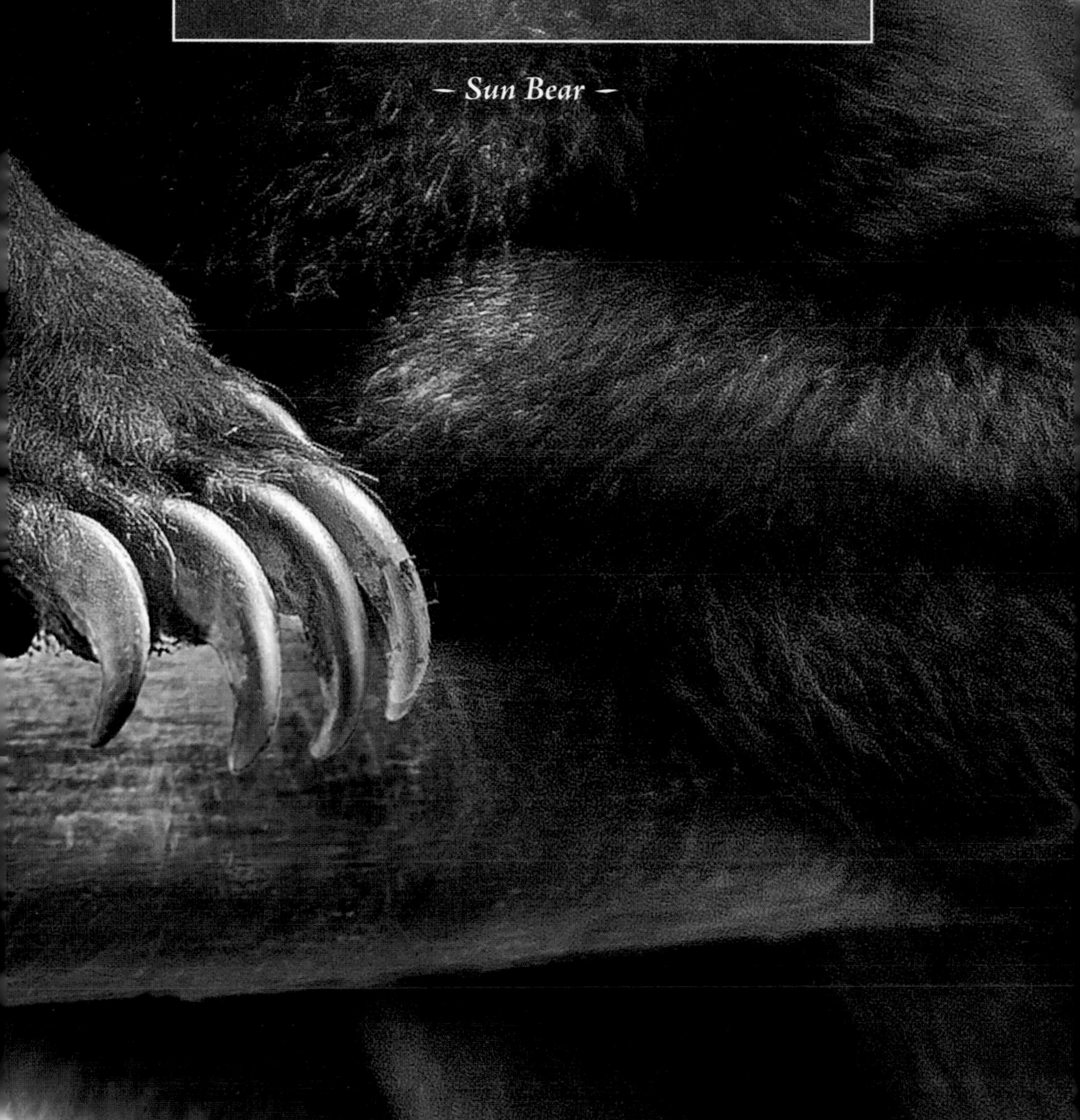

The Zoo has cared for Andean bears since 1938. Zoo researchers believe one key to the survival of the species is in conserving the critically endangered sapote tree in Peru.

Asian Passage

In Asian Passage, spreading branches of palms and ficus trees mix with towering bamboo and fragrant ginger to create a paradise for species like lion-tailed macaques, binturongs, langurs, fruit doves, rhinoceros hornbills, and porcupines. The area is designed to look like a tropical Asian forest and is rich with the sounds of birds chirping and monkeys calling. As one of the Zoo's oldest areas, Asian Passage features wide streets and exhibits that were dug into the hillsides back in the 1920s. Many of the plants have been part of the area for decades, although some were introduced later to create the lush canopy overhead. In addition to birds and primates, Asian Passage is home to several species of bears, including Bornean sun bears, brown bears, and sloth bears.

Capuchin monkeys raise the hair on their forehead to make them look bigger to rivals.

Hornbill

With long eyelashes, dark eyes, and an almost comically large, curved bill, hornbills are memorable birds. Their eyelashes are made of modified feathers. They also have a special body part atop their bill called a casque. Researchers believe this structure acts as a vibrating chamber to make the hornbills' voice louder.

Most hornbills are omnivorous, eating a combination of fruit, insects, and other small animals. They can use the tip of their bill as a finger to pluck fruit from trees or animals off the ground, and the edges of their bill are notched like a saw for grasping and tearing.

Female hornbills are dedicated mothers, sealing themselves up in the hollow of a tree for up to four months while raising their chicks. They create the seal shortly after breeding by using regurgitated food, droppings, and mud brought by the male. Once they lay their eggs, they sit on them while the male brings food through a small slit in the wall. Females keep their nest clean by dropping their waste through the opening. In some species, females also molt most of their wing and tail feathers, leaving them unable to fly for several weeks. In other species, they break out of the hollow when the chicks are half grown. The chicks then reseal themselves in the nest and are fed by both parents until they are ready to fledge.

Mission Conservation

San Diego Zoo Global has one of the country's most comprehensive collection of hornbills. Hornbills are critically endangered, threatened by habitat destruction and hunting. Many are killed for their casque, which is used for carvings and nontraditional medicines. In Sarawak, Malaysia, hornbills were hunted for their feathers until the Wildlife Conservation Society (WCS) asked ceremonial leaders and dancers if they'd be willing to use donated feathers—and the leaders agreed. Now, zoos like the San Diego Zoo and Safari Park gather shed hornbill feathers for regular shipments to Sarawak.

Porcupine

The best defense is a great offense. The quills of a porcupine are modified hairs made out of keratin. Porcupines have muscles at the base of each quill that raise them up when the animal is excited or alarmed. Like all hairs, quills do shed, and when a porcupine shakes its body, its loose quills can come off. Puncture wounds inflicted by porcupines are very serious.

Porcupines are grouped as either Old World or New World. Old World crested porcupines have back quills that can be 20 inches long. When threatened, they erect these quills, appearing two to three times bigger. At the base of their tail, they have shorter, blunt, hollow quills that they rattle to warn predators. They also charge backward into their predators or stamp their feet, growl, and grunt.

By contrast, New World porcupines—such as North American porcupines—have quills that are about four inches long that jut out in various directions when the animal feels threatened. These quills have small barbs that snag the flesh. These porcupines also lash out at predators by quickly batting at them with their quill-laden tail while chattering their teeth.

Old World porcupines spend their life on the ground, while New World porcupines spend most of their time moving through trees, using their long, curved claws for climbing.

DID YOU KNOW?

Some Indian crested porcupine quills are 20 inches long. Quills can be as dense as 150 per square inch, giving one animal as many as 30,000 quills.

Lion-Tailed Macaque

They grimace at their keepers and have been known to trash their exhibit. But lion-tailed macaques are strikingly beautiful animals and one of the most endangered species on the planet. Their wild habitat is decreasing as timber is harvested and tea, coffee, and eucalyptus plantations are established. Macaques are also hunted for food.

Lion-tailed macaques get their name from the lion-like tuft on their tail and their beautiful silvery "mane," which helps protect them from showers in their native rain forest habitat in India. These social monkeys live in troops of 10 to 20 and spend their days grooming each other, hunting for insects, and foraging for food. Macaques move through the treetops, stuffing their cheek pouches full of fruit, seeds, buds, and leaves. When they're ready to eat, they rub their cheeks with their hands to access the food.

Behind the Scenes

Lion-tailed macaques like to keep busy, so the Zoo provides enrichment for them at least three times a day. These energetic monkeys love ripping boxes and bags open to see what's inside. Their natural curiosity has led them to fiddle with and even undo parts of the exhibit: the first day on exhibit, the troop reached through the wire mesh and successfully redirected the operating sprinkler heads toward visitors!

Sun Bear

DID YOU KNOW?

Sun bears can turn in their loose skin and bite their attacker—even if their attacker has already latched on.

Sun bears are the smallest of the world's eight bear species. They get their name from the white or yellowish crescent marking on their chest, which many people think looks like the rising or setting sun. In the Malay language, sun bears are called *basindo nan tenggil*, which means "he who likes to sit high."

Sun bears like to make their home in the branches of trees. Their small size, four-inch claws, and large paws with hairless soles help them move with ease high up in the trees. And since they're nocturnal, those branches also make a nice place to build a nest for resting or sunbathing during the day.

Sun bears use their front paws to rip open trees in search of insects or sap, and their long tongue is perfectly suited for getting at honey and insects inside trees and other tight places. They also eat small birds, fruit, honey, lizards, rodents, soft parts of palm trees, and even coconuts. Unfortunately, their appetite for coconuts, oil palms, and other commercial crops have led to a lot of trouble between sun bears and humans.

Sun bears are an endangered species. It's unknown how many are left in the wild, since their secretive nature makes them hard to study. Yet researchers do know that their numbers are decreasing, due mainly to habitat loss from farming and logging, poaching, and even the pet trade.

Colobus Monkey

FIELD NOTES

Monkeys are very social animals, so it is important that they communicate well in order to get along in their large groups. They use vocalizations, facial expressions, and body movements to communicate. Staring, for instance, is a threat in monkey society. Monkeys look down or away to avoid threatening other monkeys, thus preventing fights. Grinning, or pulling the lip up to show the teeth, may seem like a smile to us. But for monkeys, this is a sign of aggression or anger.

With their thick black-and-white fur, colobus monkeys are among the world's most distinctive primates, spending most of their time in the upper part of the forest canopy. Their digestive system is similar to a cow's, with a three-part stomach that enables them to digest large amounts of leaves.

Unlike most primates, colobus monkeys do not have thumbs. Instead, they use their four fingers to grip the branches as they move quickly through the trees. They are known to leap great distances to avoid predators and use branches as springboards to jump into the air. Their mane and tail are thought to act as a parachute that slows them before they grab the next branch.

Colobus monkeys live in family groups consisting of one male, several females, and their offspring. The females take shared roles in raising the babies, helping to maintain the cohesion of the group.

Colobus monkeys defend their territory with loud vocalizations that can be heard a mile away.

Francois' Langur

Named after the Frenchman who discovered them, Francois' langurs are native to Southeast Asia. They're also known as "leaf monkeys" due to their diet, which consists primarily of leaves as well as fruit, seeds, and sometimes insects. Digesting leaves consumes a lot of energy, so these lemurs spend many hours resting. Troops consist of one male, many females, and their offspring, and can number up to 30 individuals. Francois' langurs move easily from the jungle floor to narrow perches high in the trees. Sometimes they protect themselves from extreme weather and predators by hiding out in caves.

Francois' langurs are endangered, with less than 2,000 left in the wild, but they continue to be hunted for medicinal purposes. Fortunately, protections have recently been put in place to prevent their extinction.

FIELD NOTES

Langurs may be black as adults, but they're born with a bright orange coat. Some scientists believe the orange color, which fades after three to six months, serves to remind the females in the group to help with the caretaking.

Sloth Bear

The name "sloth bear" is a bit misleading. They aren't related to sloths and they aren't slow moving. In fact, these fairly agile bears can run faster than humans. Sloth bears were named for their extremely long, thick claws and unusual teeth—features that are similar to tree sloths. Like tree sloths, they've also been seen hanging upside down on tree branches.

Sloth bears are a little unusual looking, with long, rough, unruly hair around their ears, shoulders, and neck; a pale muzzle; and a white patch of fur on their chest in the shape of a *Y, O,* or *U.* They lack hair on their nose and can open and close their nostrils as needed in order to eat their favorite foods: ants and termites. This keeps the bugs from crawling up their nose while they eat.

DID YOU KNOW?

Talk about noisy! The sucking sounds that sloth bears make while eating can be heard up to 330 feet away.

Sloth bears make a lot of noise, using their long claws to rip apart termite nests in soil, old logs, or trees. They're also noisy eaters. The large gap between their upper teeth makes the perfect space for sucking up termites, and their lips and tongue act like vacuum cleaners, creating a powerful suction and loud slurping, sucking sounds.

While sloth bears spend much of the time alone, it is believed that they actually keep the same mate each year. A nocturnal, shy species, they often sleep in caves and near rivers when available. Unfortunately, they are known as one of the most dangerous animals in central India. People go out of their way to find and kill the innocent bears. In reality, sloth bears are typically aggressive only when startled or confronted. They play an important role in their habitats as seed dispersers.

North American brown bears are called grizzly bears because the tip of their fur is white or tan, making them appear "grizzled," or streaked with gray.

BEHIND THE SCENES

The Zoo's grizzly bear brothers love to play and even sleep together in a tangled heap. The keepers work hard to provide them enrichment and keep them active. They've had their exhibit lined with sod and filled with snow, and they often drag enrichment items and palm fronds into their pool. These intelligent and rambunctious bears enjoy playing with cardboard boxes, magazines, phone books, and colorful plastic toys.

Brown Bear

Brown bears come in all sizes and shades of brown, from light cream to almost black. Adult brown bears aren't as comfortable in trees as other bear species, although cubs are encouraged to climb for safety. Like many bears, brown bears can spend up to six months deep in sleep, living off the fat their body has stored in the summer and fall. Sows even give birth during their winter sleep. By the time they emerge from their den, they've often lost up to one-third of their body weight—more if they have had cubs to nurse. Once awake, brown bears will eat anything nutritious they can find, digging for roots, tubers, insects, and sometimes hunting small prey. And each summer, coastal dwelling brown bears feast during the salmon run.

With a large hump of muscle on top of their shoulders and giant front claws, brown bears are the most powerful predators in their habitat. They have a fearsome reputation and can be more aggressive than most other bear species. However, most bears won't attack humans unless they feel threatened. Brown bears once ranged throughout the Northern Hemisphere. Due to humans' fear, heavy hunting for meat and sport, and medicinal uses, their numbers have declined. Their habitat is also being lost or fragmented at an alarming pace—the result of climate change, resource extraction, and human population growth. By altering our own habits and making conscientious changes in how we buy and use products, we can reverse these trends and save the world's bears.

Andean Bear

Native to South America, Andean bears—also known as spectacled bears—are named for the rings of white or light fur around their eyes, which can look like eyeglasses (or spectacles) against the rest of their fur. These markings often extend down the chest, giving them a unique appearance. The markings also provide their scientific name: *Tremarctos ornatus*, or decorated bears.

Andean bears are truly arboreal, building leafy platforms in the trees to feed and sleep. Because of their tropical native climate, Andean bears are active year-round. Primarily plant eaters, they dine on fruit, bromeliads, and palms. Those living in scrubland habitat are even known to eat cacti! Mostly solitary, Andean bears may gather to eat where food is plentiful. Their eating habits play an important role in forest ecology: the seeds they eat are excreted as the bears move around, spreading the seeds over long distances for the production of the next generation of fruit trees.

Very little is known about these bears in the wild, as they are shy and tend to avoid humans. Females give birth to one to three cubs in a protected, out-of-the-way den, but almost nothing is known about how they choose their den site. Researchers believe that the cubs are at least one year old when they venture out on their own.

FIELD NOTES

Andean bears are thought to use vocal communication more than almost any other bear. Their unique vocalizations are quite "un-bear-like": a shrill screech and a soft, purring sound. Mother bears may use different vocalizations to communicate with their cubs.

Camellia

Its delicate beauty and range of blossom colors have made the camellia a popular garden plant for centuries. This relative of the tea plant is native to China, Japan, and Indo-China. It was first introduced to Europe in the early 1700s and arrived in America in the late 1700s, where it became a staple of gardens in the South. Most of the camellias seen today are hybrids that come in a variety of forms, from peony and formal double to anemone-centered and single.

CATCH A GLIMPSE

Look for colorful camellias along the path that leads from Owens Aviary to Center Street. To see the type that tea is made from, *Camelia sinensis*, look around the giant panda enclosures. Late winter is the best time to see camellias in full bloom.

Camellias grow best in partial shade. Young plants particularly thrive in the shade of tall trees.

Lost Forest

– Orangutan –

Mandrills can live in troops of up to 10 animals. "Super troops" of several hundred, however, may gather when food is readily available.

Lost Forest

Lost Forest is the largest zone at the San Diego Zoo. It starts at the entrance to the Zoo and winds through a lush forest past some of the Zoo's most popular animals, including tigers, otters, hippos, and orangutans. For bird lovers, the zone includes three walk-through aviaries—Scripps Aviary, Owens Aviary, and Parker Aviary—as well as the Australasian aviaries: 36 exhibits with diverse, feathered birds from Australia, New Guinea, and Tasmania. Lost Forest also features one of the Zoo's hidden jewels: Fern Canyon Trail, a lush refuge with ferns, waterfalls, orchids, and palms. Throughout the zone, African mahogany trees, palms, and vines form a verdant canopy, re-creating the tropical rain forests that are home to so many rare and endangered animals.

The Andean cock-of-the-rock, the national bird of Peru, is known for its elaborate courtship displays.

Mandrill

Endangered mandrills are one of the largest—and most unique looking—monkey species in the world. Males have impressive coloration, with thick purple-and-blue ridges along their nose, a red nose and lips, a golden beard, and bright hues on their rear end. Adult male mandrills with the brightest and most distinctive colors are thought to be the most attractive to females. The leader, or dominant male of each troop, has the boldest, brightest colors.

Mandrills have large cheek pouches that they can stuff full of food to eat later. Sometimes they take their goodies to a safer location before enjoying them. They spend most of their time on the ground, foraging for seeds, nuts, fruits, leaves, roots, fungi, and small animals. They do climb trees, though, selecting a different tree to sleep in each evening.

Mandrills are most closely related to and share a habitat with drills, one of the most critically endangered primate species in Africa. Conservation organizations are working to protect their habitat from illegal logging and the illegal bushmeat trade. The bushmeat trade is the biggest threat to both drills and mandrills. By protecting their habitat, both species can be saved.

FIELD NOTES

Mandrills communicate through vocalizations, body language, and possibly scent markings. Sometimes they shake their head and "grin" widely to show their enormous canine teeth, which can be over two inches long. This may appear scary to humans, but it's usually a friendly gesture within the mandrill community.

Siamang

DID YOU KNOW?

Siamangs are one of the few primates known to form monogamous pairs. Paired males and females sing duets to one another, with each pair creating a unique song.

The largest species of gibbons, siamangs are well suited for life in the treetops. Unlike great apes, siamangs do not build nests; they sleep sitting upright on branches in the higher parts of trees. Siamangs have extra-long arms that can cover up to 10 feet in a single swing. They also walk along branches with their arms outstretched. On the rare occasions they choose ground travel, they walk on two legs, holding their arms over their head for balance.

Siamang pairs usually stay together for life. A siamang family group consists of one adult male and one adult female, along with two or three immature offspring two or three years apart in age. Fathers help raise the babies and take over the youngsters' daily care when they're about one year old. Young siamangs stay with their family for approximately five to seven years (even up to nine) before starting their own family group.

The loudest of all the gibbons, siamangs can be heard up to two miles away. Their booms and barks are made louder by their inflatable throat sac. These calls are used primarily for claiming territory, which can be as large as 50 acres. First thing in the morning, adult females start their territorial hooting, which their family members join. The noisy warning can last 30 minutes.

Newly independent siamangs may spend several years calling through the forest, looking for a mate.

Native to Indonesia, Malaysia, and Thailand, fire-tufted barbets have a unique call that sounds like the buzzing of a cicada.

BEHIND THE SCENES

When birds are introduced to an aviary at the Zoo, they spend a few days in an acclimation cage—sometimes called a "howdy cage." There, the resident birds can check out the newcomers through the mesh, and newcomers can get the lay of the land. Typically, the birds are not aggressive toward one another, but this helps to minimize stress for the new additions.

Owens Aviary

Built in 1937, the Owens Aviary is one of the world's largest walk-through avian encounters. Towering nearly 82 feet, this free-flight aviary features more than 200 individual birds representing 45 species native to Southeast Asia.

The aviary begins at treetop level and winds down a pathway through a Southeast Asian forest canopy filled with foliage, flowers, and ferns. Rocky waterfalls spill into pools, and comb-crested jacanas stroll across the floating lily pads. Visitors might see a rare woodpecker chiseling through the bark of a tree or the courtship ritual of the Argus pheasant at the forest-floor level. Several feeding trays are located close to the walkway railing, providing great spots to see and photograph the birds.

Harpy Eagle

Harpy eagles are named after the predatory "frightful flying creatures" of Greek mythology. Ranging from Mexico to northern Argentina, they are among the heaviest and most powerful of flying birds. "Harpies" have legs as thick as a small child's wrist, a wingspan of six and a half feet, and five-inch-long back talons that are larger than a grizzly bear's claws.

With their deadly talons, harpies can exert several hundred pounds of pressure to kill their prey. They fly below the forest canopy, snatching up monkeys and sloths that can weigh up to 17 pounds. Harpies are capable of flying 50 miles per hour, and they can fly almost straight up, too, enabling them to attack prey from below as well as above. These birds are patient hunters, perching silently—sometimes for up to 23 hours—as they wait to catch unsuspecting prey. With their excellent vision, they can see something less than one inch in size from almost 220 yards away.

Harpy eagles are monogamous and may mate for life. Their nest is about four feet thick and five feet across—large enough to hold a human! As parents, they fiercely defend their eggs and young. Females lay one or two eggs, reproducing every two to three years. The first eaglet to hatch gets all the attention and is more likely to survive, while the other dies from lack of incubation. This second egg acts as an insurance policy in case there is something wrong with the first egg. Once mature, chicks sometimes return to nest in their "home tree." Given their years of dedicated parenting, harpy pairs may not raise many offspring in their lifetime.

Mission Conservation

Harpy eagles need several square miles of undisturbed forest to thrive. Unfortunately, years of logging, destruction of nesting sites, poaching, and hunting have eliminated many of these birds—and since harpy parents raise just one eaglet every two years, it is difficult for populations to recover. However, captive breeding can help build populations by providing protected nesting sites and hand rearing of some chicks, allowing the parents to lay another egg and raise another chick. The San Diego Zoo was the first facility in North America to hatch and successfully rear a harpy eagle chick in 1992. Since then, 15 harpies have hatched at the Zoo and two of the offspring have been released in Panama.

Okapi

With their white-and-black striped hindquarters and front legs, okapis (pronounced "oh-COP-ee") look like they are related to zebras. In reality, they are the only living relatives of the giraffe. Like giraffes, okapis have large, upright ears that catch even slight sounds. They use their long, dark prehensile tongue to strip buds and young leaves from the understory brush of the rain forest. They eat 40 to 65 pounds of leaves, twigs, and fruit each day. Baby okapis can stand up within 30 minutes of birth. Mothers hide their newborn calves in one spot, returning regularly to nurse.

Okapis are difficult to find in the wild; they are secretive by nature and their natural habitat is the Ituri Forest, a dense rain forest in Central Africa. Their highly developed hearing alerts them to run when they hear humans in the distance. While natives of the Ituri Forest were familiar with okapis, the Western world did not know of their existence until 1900.

DID YOU KNOW?

Okapis use their long, dark tongue to clean their eyes and ears. Their ears move independently so they can listen for sounds both in front and behind.

Orangutan

Orangutans live in tropical and swamp forests on the Southeast Asian islands of Borneo and Sumatra. These shaggy red apes are the largest arboreal mammals and the only great apes found in Asia. Orangutans spend most of their life in the trees, swinging from branch to branch in search of fruit. Females usually have their first baby around age 15. Babies hold tight to their mother's belly as she swings through the forest. When they're older and better at balancing, they ride piggyback. Youngsters stay with their mother until they're about eight years old, the longest childhood of the great apes. From their mother, they learn to find fruit in the rain forest, build nests for sleeping at night, and other survival techniques.

While other great apes are usually found in groups, orangutans are more solitary. However, when mothers and their youngsters encounter each other in their overlapping ranges, they peacefully feed together and watch their youngsters play. Adult males allow females to feed in their range and will tolerate nondominant males that lack cheekpads. If a mature "cheeked male" swings into a territory occupied by a more dominant male, the dominant male gives a booming roar called a "long call" to scare the intruder off. Males can be aggressive, sometimes charging each other and breaking branches—or even grabbing and biting one another until one of them gives up and runs away.

Mission Conservation

These highly intelligent red apes are now extinct in much of Asia. Most of their forest habitat has been destroyed to make room for palm oil plantations, and poachers often kill orangutan mothers and sell their young to the illegal pet trade. San Diego Zoo Global participates in the Association of Zoos and Aquariums (AZA) Species Survival Plan for Sumatran orangutans and provides funding to the Association of Zoos and Aquarium's Ape Taxon Advisory Group's Conservation Initiative. In addition, Zoo researchers have aided conservation efforts by identifying chromosomal differences between Bornean and Sumatran orangutans, a key factor in maintaining genetic integrity and diversity.

Orangutan youngsters are much more active and social than their parents. They weigh about four pounds at birth.

Allen's Swamp Monkey

As their name implies, Allen's swamp monkeys live near water and are good swimmers. They have webbed fingers and toes and have been known to dive into the water to escape predators like snakes, eagles, and bonobos. They also "fish" by laying leaves or grass on top of the water and grabbing the fish that hide underneath.

These monkeys may live in groups of around 40 individuals, which are then divided into subgroups of two to six monkeys who work together to forage for food. The groups' sleeping sites are usually located near water. Individuals will groom each other to remove dead skin and parasites. This kind of behavior also reinforces bonds.

CATCH A GLIMPSE

Watch for the juvenile swamp monkeys, who like to play with the African spotted-necked otters, touching the otters' tail so the otters will chase them. Some of these spunky monkeys have even been seen doing cannonballs into the pool!

Like all guenons, Allen's swamp monkeys use their cheek pouches to store extra food while foraging.

River Hippo

Adapted for semiaquatic life, river hippos are found in slow moving rivers and lakes in Africa. River hippos have eyes, ears, and nostrils on top of their head, enabling them to breathe, hear, and see what's happening above the water without exposing themselves. Their eyes have special nictitating membranes that act as built-in goggles, allowing them to see underwater, and they can hold their breath for up to 30 minutes. They also sleep underwater—a reflex action allows them to bob up, take a breath, and sink again without waking up.

Mature male river hippos can weigh up to four tons. They eat about 1 percent of their body weight each day—predominantly grass. At dusk, they leave their watery hangout and travel for miles, flattening plants in their path and using their thick lips to rip grass from the ground.

Hippos face many perils such as disease and drought. Juveniles are hunted by crocodiles, lions, hyenas, and leopards—and adults are even threats to each other. After the 1989 ban on elephant ivory, demand for hippo ivory sharply increased. Their canines are made of the same material as elephants' tusks, but hippo ivory is slightly softer and easier to carve than elephant ivory, which makes it even more appealing.

If hippos were to disappear, the effect on their habitat would be catastrophic. The large amount of waste that hippos produce fertilizes the African ecosystem, and many species of fish eat the dung and feed on the small parasites that live on hippos' skin.

FIELD NOTES

River hippos are responsible for more human deaths each year than any other African mammal. They can reach speeds of 20 miles per hour on land and move quickly through the water. They are territorial and protective. Bulls stake a claim to a stretch of river and all the females in it. Females, especially mothers with calves, can be dangerous and unpredictable. They will react aggressively if crossing paths with boaters.

Researchers estimate that there are no more than 5,000 tigers remaining in the world. Fortunately, many countries have passed laws against killing tigers, and international projects now exist that protect their habitat.

Behind the Scenes

The tigers at the Zoo receive basic behavioral training to assist in their daily care. They "rise up" on their hind legs to allow staff to inspect their belly and paws, open their mouth so keepers can ensure their teeth are healthy, and allow blood draws from the base of their tail. The animals' calm cooperation allows them to avoid being tranquilized for routine health care.

Malayan Tiger

Tigers have large, strong front paws to bring down their prey. Their claws can be pulled inside while they walk, which helps keep them sharp. All tigers have a uniquely patterned coat, and those who observe tigers can identify individuals by their particular stripes. As stalk-and-ambush hunters, tigers are camouflaged in tall grass by their orange, black, and white stripes, which break up their outline as they stealthily approach prey.

Tigers have a white spot on the back of each ear. These spots may help them find one another, or they may be a way for mothers and cubs to keep one another in sight in the dense forest undergrowth. Some researchers believe the spots are designed to resemble staring eyes to scare predators that may be behind them.

All six species of tigers are endangered due to habitat loss and hunting for their body parts; three other subspecies are already extinct. Despite international outcry and the lack of scientific endorsement, tiger bones and organs are still used in folk medicine. Tigers have lost 93 percent of their historic range due to increasing human populations. As opportunistic hunters, these solitary cats each require a significant home range to ensure they meet an estimated kill of 50 large prey animals per year.

Steller's Sea-Eagle

FIELD NOTES

Steller's sea-eagles build nests, or aeries, up to 100 feet off the ground in dead or open topped trees or on rocky cliffs. These open sites give the birds easy access to and from their nest. Pairs typically return to their nest each year, adding a little more to it to prepare for the season's clutch. One aerie can be six to eight feet wide and weigh hundreds of pounds.

Named for the noted 18th-century zoologist and explorer Georg Wilhelm Steller, Steller's sea-eagles are only found on the northeastern Russian coast and in North Korea and Japan. Very little is known about these birds of prey due to their remote habitat. They live along tree-lined river plains and rocky coastlines. These birds mostly prey on trout and salmon, and they take full advantage of the annual salmon run to gorge themselves.

As of 2006, the world's population was estimated at 5,000 birds, but it is slowly decreasing. While the eagles are legally protected in Russia and Japan, they are losing their habitat due to hydroelectric power projects and logging. In addition, chemicals from local industries have contaminated the rivers where they fish. Overfishing by humans in Japanese waters has led the eagles to scavenge on sika deer remains left by hunters. Eating carrion filled with lead bullets has had devastating effects on the eagle population, causing Japan's Hokkaido Island to outlaw lead ammunition.

The San Diego Zoo and Natural Research Ltd are studying the movements of young Steller's sea-eagles in their native habitat in hopes of protecting the species in the wild. The Zoo has also loaned pairs of the sea-eagles to four other zoos in the United States.

Wolf's Monkey

Wolf's monkeys are social, mischievous primates, native to the rain forests of the Congo Basin in Central Africa. They're not related to wolves, though—these monkeys are named after the person who first described them for science.

Wolf's monkeys belong to a group of medium-sized monkeys called guenons. Unlike other tree-dwelling and leaf-eating monkeys, guenons have more varied diets, eating fruit, seeds, young leaves, insects, some reptiles, and occasionally small mammals. They store extra food in special cheek pouches to eat later. These pouches can hold almost as much as their stomach. Wolf's monkeys also often feed on nectar, making them important pollinators in the forest.

Most guenons live in large groups with other guenon species and primates like mangabeys, colobus monkeys, and Allen's swamp monkeys. This safety in numbers helps them spot predators like chimpanzees, leopards, and eagles. Guenons use a variety of calls to stay in touch while they leap from branch to branch foraging. Scientists have recorded seven different sounds, including a grunting call they use while looking for food. (Their grunting increases when they find insects!) They also communicate with several gestures and facial expressions: an open mouth showing sharp teeth, closed eyelids, or a movement of their head or tail.

BEHIND THE SCENES

During the day, the Zoo's Wolf's monkeys share a mixed-species exhibit with the pygmy hippos, where they dash around, blithely swinging from vine to branch to treetop. One of the monkeys, a female, even likes to jump on the back of one of the hippos for a ride. (The hippo doesn't seem to mind.) At night, the monkeys go to their bedrooms, while the more nocturnal pygmy hippos get free run of the area.

Pygmy Hippo

At first glance, pygmy hippopotamuses look like miniature versions of their larger, better known relatives, river hippopotamuses. But pygmy hippos differ in both behavior and physical characteristics. They tip the scales at a delicate 350 to 600 pounds, versus three and a half tons for river hippos. They are also much more rare than river hippos. Found only in the interior forests in parts of West Africa, these shy, nocturnal herbivores eluded Western science until 1840.

Pygmy hippos are usually found alone or in pairs. They rest well hidden in swamps, wallows, or rivers during the day and leave the water to feed on land for a few hours in the cool of the night.

Although they can make noises—from low grunts to high-pitched squeaks—they are usually silent, relying mostly on body language to communicate. With their cavernous mouth and formidable teeth and tusks, they can scare away potential enemies with a simple "yawn." They also protect themselves using other threatening behaviors such as rearing, lunging, scooping water with their mouth, and shaking their head. Unlike their larger relatives, pygmy hippos are shy and would prefer to flee rather than fight.

DID YOU KNOW?

All hippos have mucous glands that produce an oily red secretion called "blood sweat" that moistens and protects their skin from the sun. This slimy substance is believed to also help heal their nicks and scrapes.

African Mahogany

African mahogany trees are native to the moist, evergreen forests of tropical Africa and Madagascar.

In Africa, these trees flower at the end of the dry season, bearing clusters of tiny white blossoms that scent the air. The fruit transforms into a woody, round seedpod over the course of a year. When the seedpods break open, hundreds of "winged" seeds are released. The flattened edges catch the air in a way that causes the seeds to spiral away from the parent tree as they descend. In this way, the seeds have a better chance of landing in a spot that will allow them to germinate and grow.

CATCH A GLIMPSE

The Zoo's African mahogany trees were carefully nurtured from seedlings in a staging area before putting down roots along Monkey Trail. Here, as in their native habitat, the mahoganies contribute to a lush, varied canopy. Look up and see if you can spot new growth, which is usually red.

Western Lowland Gorilla

Gorillas are the largest of all primates—the group of animals that includes apes, monkeys, and prosimians such as lemurs. These peaceful, family-oriented omnivores love to eat. Adult males consume up to 40 pounds of leaves, bark, stems, fruit, seeds, and roots each day, and even ants and termites when they are available. Their large stomach can hold the bulky food, and their strong jaws help them chew tough stems. The Zoo's gorillas are offered a variety of seasonal fruit and vegetables, plus banana and ficus leaves.

Gorillas are social animals that live in a group called a troop. Each troop is made up of five or more gorillas, led by a mature, experienced male known as a silverback. The silverback is responsible for the safety and well-being of the members of his troop. He makes all the decisions, such as where the troop will travel each day, when they will stop to eat or rest, and where they will spend the night. In Africa, gorillas usually don't stay in the same place for more than a day. Every morning, the silverback leads his troop to a new area where food is plentiful. After a morning of munching, the adults gather leaves, twigs, and branches to make day nests for resting while the youngsters play. Then the gorillas eat again until bedtime and make another nest for a good night's sleep.

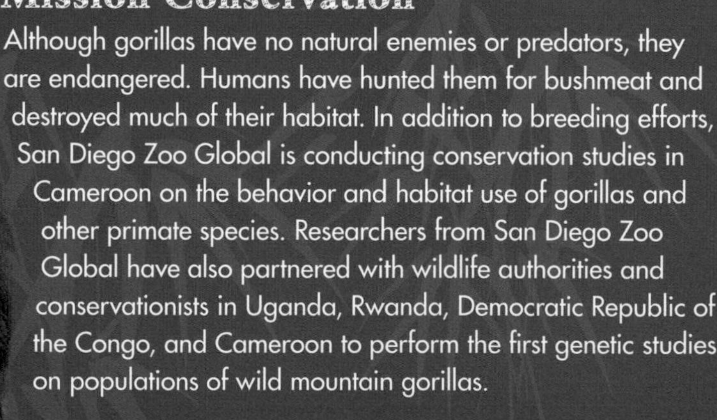

Mission Conservation

Although gorillas have no natural enemies or predators, they are endangered. Humans have hunted them for bushmeat and destroyed much of their habitat. In addition to breeding efforts, San Diego Zoo Global is conducting conservation studies in Cameroon on the behavior and habitat use of gorillas and other primate species. Researchers from San Diego Zoo Global have also partnered with wildlife authorities and conservationists in Uganda, Rwanda, Democratic Republic of the Congo, and Cameroon to perform the first genetic studies on populations of wild mountain gorillas.

No two gorilla noses are alike. In the wild, researchers take close-up photos of each gorilla's face to help identify individuals.

Malayan Tapir

DID YOU KNOW?

Malayan tapirs are the most primitive large mammals in the world. They have remained unchanged for millions of years. Fossils of tapir ancestors have been found on every continent except Antarctica.

They may look like pigs or anteaters, but tapirs are neither. Tapirs are most closely related to horses and rhinos, since each one of their toes has a separate hoof. Their nose and upper lip are combined into a flexible snout like an elephant's trunk (just not as long!), which they use to grasp and pull plant material into their mouth. In addition, their body is teardrop shaped: tapered in the front and wider at the rear, designed to walk through thick vegetation.

Tapirs are good swimmers, their snout functioning like a snorkel when they're paddling just under the surface of the water. Tapirs also walk along the bottom of lakes or rivers, much like hippos. They can stay underwater for several minutes. Even youngsters can swim when they're just a few days old.

Malayan tapirs are found in Thailand, Burma, Indonesia, and Malaysia. They have a distinctive coat pattern, with black in the front and white in the back. This acts as camouflage by breaking up the body outline in the shade of the forest. Calves are born with a white spotted and stripped coat. Tigers are their natural predators.

Tapirs make a whistle that sounds like car brakes coming to a screeching halt. They also snort and stamp their feet when preparing to defend themselves.

Spot-Necked Otter

Named for the white markings on their neck and chest, spot-necked otters inhabit the fresh waters of Central and Southern Africa. They most often live in lakes, where they eat fish (and sometimes crabs and frogs). Some spot-necked otters have been known to eat fish as large as themselves!

Otters are part of the weasel family. Their sleek, streamlined body is perfect for diving and swimming. They have a long, slightly flattened tail that moves sideways to propel them through the water, and back feet that act like rudders to steer. Almost all otters have webbed feet, and they can close off their ears and nose as they swim underwater. They can stay submerged for about five minutes.

Otters have two layers of fur to help them stay warm: a dense undercoat that traps air and a topcoat of longer, waterproof guard hairs. Keeping their fur in good condition is important, so otters spend a lot of time grooming. In fact, if their fur becomes matted with something like oil, it can damage their ability to hunt for food and stay warm.

FIELD NOTES

Otters are energetic and playful, intelligent and curious. They're always busy hunting, investigating, or playing with something. They like to throw and bounce things, wrestle, twirl, and chase their tail. They also play games of "tag" and chase each other, both in the water and on the ground.

Venus flytraps are often difficult to grow but can thrive for up to 30 years if cultivated in the right conditions.

Bog Garden

Bogs are nearly sterile, acidic, low-oxygen environments that feature one of nature's most inspired adaptations: carnivorous plants. In these difficult environments, plants have adapted to lure, trap, and digest insects and other small animals in order to get necessary nutrition.

Carnivorous plants produce flowers in the spring to ensure pollination by the first insect visitors. Any insects that come along after that, however, are not so lucky. They entice their victims with bright colors and scents like sweet nectar or the stench of animal carcasses. Then they use a variety of techniques—from narcotic secretions to adhesive liquids to leaves that snap shut—to trap the creatures. Digestive enzymes break down their prey.

Located at the end of Monkey Trail, the Zoo's Bog Garden features Venus flytraps, pitcher plants, sundews, and bladderworts. To provide these plants with their natural environment, the area is kept swampy with deionized water that is filtered to eliminate the salts and minerals found in San Diego's water supply. The garden is completely saturated every four days to mimic the rainy conditions the plants are used to.

Discovery Outpost

– Macaw –

Giant tortoises eat prickly pear cactus and fruit as well as flowers, water ferns, leaves, and grasses.

Behind the Scenes

All the giant tortoises at the Zoo have a different personality. Some are shy while others are interactive and affectionate. Some love being petted and will stretch up for a neck rub or eat right out of a keeper's hand. One tortoise knows how to give the keeper high fives and can shake hands.

Discovery Outpost

Anchoring the east end of the Zoo, Discovery Outpost includes a host of special exhibits, including the Reptile House and Reptile Walk, the Hummingbird Aviary and Insect House, and the Children's Zoo. The Children's Zoo has a Petting Paddock full of gentle goats and wooly sheep, a playground, and a 4-D theater, where characters from the animated film *Rio* take guests on an adventure through a colorful rain forest. There's also an animal nursery with large viewing windows, where at-risk animal babies are bottle-fed by their keepers; exhibits featuring animals like macaws, meerkats, miniature horses, and mice; and opportunities to meet some of the Zoo's animal ambassadors up close. Guests might be able to feel the quills of a hedgehog or listen to how quietly a horned owl can flap its wings.

About 25 percent of a naked mole-rat's muscle mass is in its jaws.

Reptile House

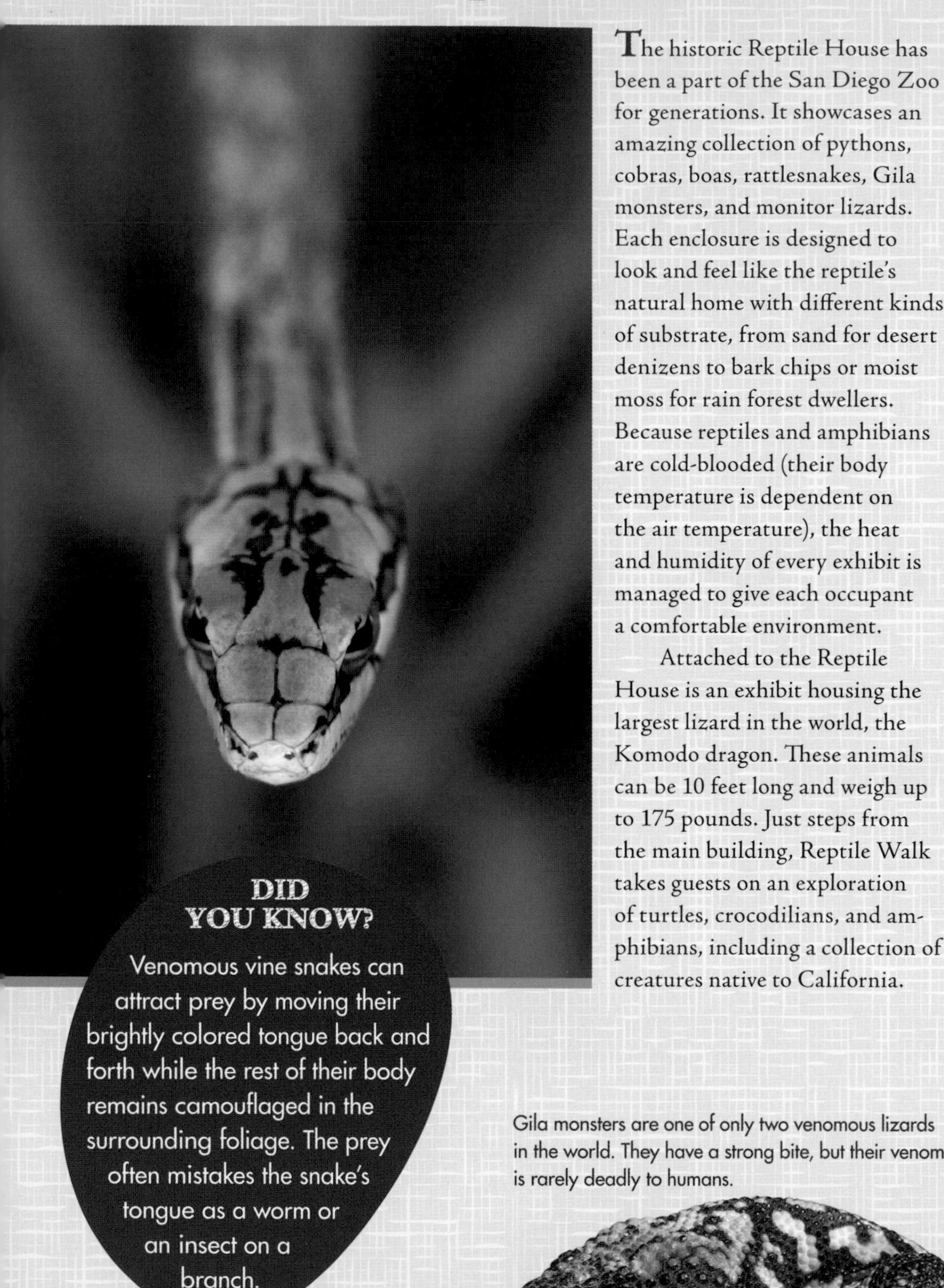

The historic Reptile House has been a part of the San Diego Zoo for generations. It showcases an amazing collection of pythons, cobras, boas, rattlesnakes, Gila monsters, and monitor lizards. Each enclosure is designed to look and feel like the reptile's natural home with different kinds of substrate, from sand for desert denizens to bark chips or moist moss for rain forest dwellers. Because reptiles and amphibians are cold-blooded (their body temperature is dependent on the air temperature), the heat and humidity of every exhibit is managed to give each occupant a comfortable environment.

Attached to the Reptile House is an exhibit housing the largest lizard in the world, the Komodo dragon. These animals can be 10 feet long and weigh up to 175 pounds. Just steps from the main building, Reptile Walk takes guests on an exploration of turtles, crocodilians, and amphibians, including a collection of creatures native to California.

DID YOU KNOW?

Venomous vine snakes can attract prey by moving their brightly colored tongue back and forth while the rest of their body remains camouflaged in the surrounding foliage. The prey often mistakes the snake's tongue as a worm or an insect on a branch.

Gila monsters are one of only two venomous lizards in the world. They have a strong bite, but their venom is rarely deadly to humans.

Unlike other reptiles, tuataras are able to stay active in colder weather. These nocturnal animals can also go an hour without breathing.

BEHIND THE SCENES

Off exhibit near the Reptile House are some of the Zoo's most rare animals—tuataras, lizards native to the islands of New Zealand. Tuataras are the only survivors of an ancient group of reptiles that date back 225 million years, when dinosaurs roamed the earth. Scientists believe these endangered animals can live to be nearly 100 years old in the wild. The Zoo has the only collection of Brothers Island tuataras in North America and is working to breed them.

King Cobra

Found throughout India, southern China, and Southeast Asia, king cobras are the largest venomous snakes in the world, reaching more than 18 feet long. Like all cobras, king cobras kill their prey by injecting a neurotoxin through their fangs that stops the victim's heart within minutes. Cobras hunt at dawn and dusk, using their forked tongue to pick up odor particles from the ground and passing those particles over a special smelling organ on the roof of their mouth called the Jacobson's organ. King cobras primarily eat other snakes—even those of their own kind—and can go for days or even months without eating, depending on the size of their last meal.

King cobras have special muscles and ribs in their neck that spread out to form a "hood" when they feel threatened. This makes them look bigger than they really are and may help scare away predators. They also make a loud hiss and rear up as high as one-third of their body length—sometimes reaching six feet tall.

DID YOU KNOW?

The sight of a large cobra reared up will stop an elephant in its tracks. A king cobra can deliver enough venom in a single bite to kill an elephant.

Fijian Banded Iguana

The iguana family includes some of the largest lizards found in the Americas. Like other reptiles, iguanas are cold-blooded, egg-laying animals with an excellent ability to adapt to their environment. They vary greatly in size, color, and behavior. In fact, some species look and act so differently that they seem like members of another family.

Fijian banded iguanas are among the most rare—and striking—creatures in the world. Named for the bands of light blue and green across their body, these animals are rarely seen in the wild. They have a long tail and toes and sharp claws that enable them to spend most of their time in the trees. Males are highly territorial, even toward females at times. They can reach up to two feet long.

Mission Conservation

The San Diego Zoo has the largest and most successful colony of endangered Fijian banded iguanas outside of Fiji. More than 100 have hatched at the Zoo since 1965. Fijian banded iguanas are particularly at risk because of their limited range: they live on only a few islands in the Fijian archipelago. To take steps to prevent this species' extinction, San Diego Zoo Global, in cooperation with the Fiji Iguana Species Survival Plan, created the Fiji Iguana Conservation Fund to support much needed conservation projects in Fiji.

Komodo Dragon

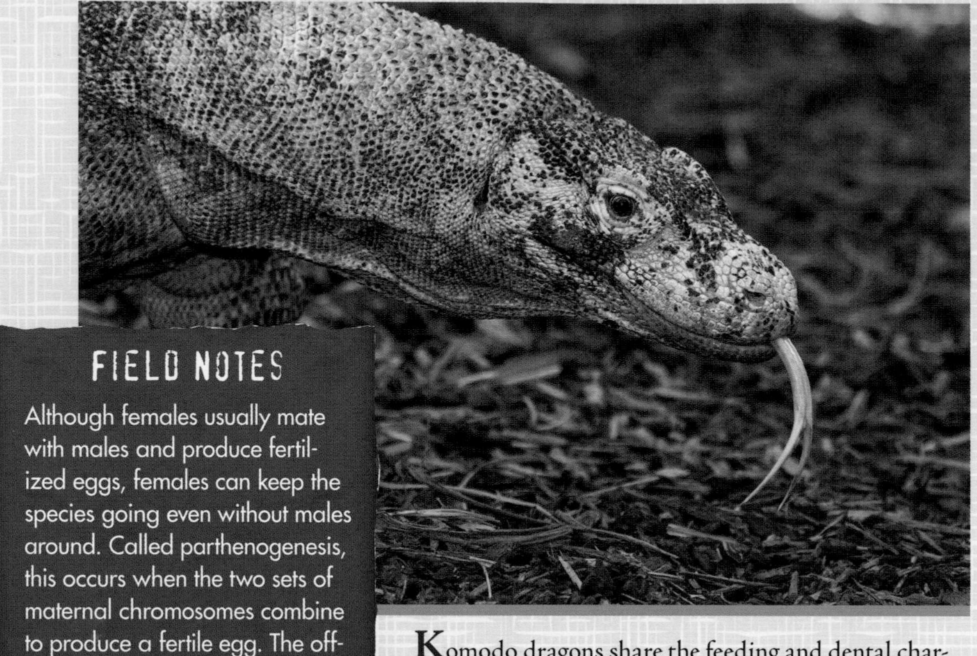

FIELD NOTES

Although females usually mate with males and produce fertilized eggs, females can keep the species going even without males around. Called parthenogenesis, this occurs when the two sets of maternal chromosomes combine to produce a fertile egg. The offspring are always male. They're also genetically distinct, so once they mature, they can mate with their mother and start a new generation of dragons.

Komodo dragons share the feeding and dental characteristics of dinosaurs, saber-toothed cats, and sharks. Their ancestors date back more than 100 million years. Komodo dragons were probably first sighted by Chinese sailors as early as the second century AD, when they navigated the waters of Indonesia. The sailors returned home with tales of huge creatures with long claws, a forked tongue, venomous breath, and scaly skin.

The top predators in their island ecosystem, Komodo dragons can reach up to 10 feet long and weigh more than 170 pounds. They swim, dive, climb trees, jump, dig, and run swiftly. These giant reptiles ambush their prey, darting out to bite their victims with slashing and tearing teeth. After the initial attack, they often retreat, letting their prey escape—but their bite has already introduced a toxic combination of lethal bacteria from the saliva and an anticoagulant from the venom. Within a few days, the victim is dead, and the dragons only have to track it down using their long, forked yellow tongue, which "smells" where the animal is waiting.

During these attacks, the lizards' serrated teeth may break off, but another set of teeth will replace them. This occurs up to five times in their life.

Komodo dragons are known to be cannibalistic. Adults often eat young lizards that have not sought the safety of the trees.

Galápagos Tortoise

Native to the Galápagos Islands of Ecuador, Galápagos tortoises can reach six feet long and weigh more than 500 pounds. Their shell looks solid, but it's actually lightweight—made of honeycomb structures that hold small air chambers. Their lungs are located under the dome of their shell. This is why tortoises can be in trouble if they get turned over—the weight of their body can crush their lungs!

Galápagos tortoises are well adapted to their environment. To keep warm on cool nights, they sleep partially submerged in mud, water, or brush. They also retain a lot of water, which allows them to survive the long dry season. In fact, they can go without eating or drinking for up to a year.

Galápagos tortoises spend much of their time grazing, sunbathing, wallowing in puddles, and ambling along at 0.16 miles per hour. When fights break out among males, they simply face each other, open their mouth, and stretch their head as high as they can. The male with the highest head wins, even if he's much smaller overall.

Giant tortoises reach maturity at around 20 to 25 years old. After breeding, females journey several miles to reach nesting areas, laying large, hard-shelled eggs in holes about 12 inches deep and covering them with sand. If the nest temperature is low, more males will hatch; if it is high, more females will hatch. Hatchlings weigh just a few ounces at birth.

The San Diego Zoo has a long history of contributing to tortoise conservation. The Zoo received its first Galápagos tortoise in 1928 and had its first hatchling in 1961—the first ever recorded in a zoo. In 1969, the Zoo became a partner with the Charles Darwin Research Station on the Galápagos Islands. Today, it has one of the largest herds of Galápagos tortoises in the world.

CATCH A GLIMPSE

Many of the Zoo's tortoises have been here since 1928, making them well over 100 years old. They can be identified by the numbers painted on their shell. Males have white numbers and females have red. The original tortoises are white 5, 6, 7, 23, and 25 and red 4, 5, 7, 8, and 9.

A tortoise's shell is attached to its ribs. Unlike the misconception shown in some cartoons, a tortoise cannot "walk out" of its shell.

Gharial

Gharials are one of the longest species of crocodile. They are noted for their long, narrow snout, which aids them in fishing, allowing them to strike rapidly in the water. Males grow a conspicuous, rounded bulge made of cartilage at the end of their snout that attracts females. This bulb, and the gharial's long snout, gives this species its name. (In Hindi, a *ghara* is a clay pot with a long neck.)

Gharials fold their limbs next to their body while swimming, using their tail to propel them through the water. Unlike other crocodile species, they do not move well on land and usually slide along the ground on their belly. They typically come out of the water only to bask in the sun or to nest.

Once widespread throughout India and its neighboring countries, gharials are critically endangered. Their population has declined approximately 98 percent since 1946, the result of overhunting for skins and trophies, egg collection for consumption, use in native medicines, and killing by fishermen. The last remaining refuge for gharials is the National Chambal Sanctuary along the Chambal River. There, conservationists are protecting nests from predators and raising hatchlings in captivity until they can safely be released into the wild.

Mission Conservation

In 2012, the San Diego Zoo received 12 juvenile gharials as part of a breeding assurance colony from the Madras Crocodile Bank Trust and Centre for Herpetology (MCBT). MCBT is one of the largest reptile zoos in the world and one of the oldest nongovernment environmental organizations in Asia. These gharials have been a significant addition to the Species Survival Plan. San Diego Zoo Global also manages the Gharial Conservation Fund, which supports research and conservation efforts in the field.

Chinese Alligator

One of only two alligator species in the world, Chinese alligators travel the swamps, ponds, and slow-moving streams around the Yangtze River, mostly feeding on snails and mussels that they crush with their short, blunt teeth. Chinese alligators reach just five feet long—much smaller than American alligators, which can be up to 20 feet long. From nose to tail, belly to back, these alligators are protected by hard scales under the skin. Even their eyelids have bony plates.

Chinese alligators are critically endangered with fewer than 130 in the wild. Much of their habitat has been altered for agriculture, leading to increased conflict with humans. They sometimes hunt ducks on farms and dig into the sides of irrigation ditches, causing the ditches to collapse. The Chinese government has been working to help farmers find ways to live with the alligators. A breeding center has been established there, and some alligators are being introduced back into the wild, where they are monitored carefully. Many others are being bred in wildlife refuges and zoos. However, survival outside of managed care can only happen if their wetland habitats are protected.

DID YOU KNOW?

Crocodilians—the group that includes alligators—have such powerful senses that they can hear their juveniles calling from inside their eggs.

With their pink and white coloring and petal-shaped body, orchid mantises easily lure unsuspecting prey.

Insect House

The Insect House is full of terrariums set into walls that showcase stick insects, roaches, beetles, scorpions, spiders, and other spineless wonders. Inside each terrarium, the temperature and humidity are monitored and adjusted to suit the preferences of the individual species. Full-spectrum lights illuminate each terrarium, providing the animals with the same range of light waves they would get naturally from the sun. The Insect House even includes a beehive enclosed in glass. A clear tube allows the bees to leave the hive—and the building—to gather nectar and pollen from the flowers outside. Part of the wall around the outer opening is painted bright yellow to help the bees find the entrance as they return home.

The Insect House also features a colony of leafcutter ants. Leafcutter ants have complex, disciplined colonies in which they work together to simulate the functions of a whole body, including organs like the brain and functions like reproduction. It all happens organically yet efficiently, without leaders. Leafcutters rely chiefly on chemical signals called pheromones. Their pheromones enable them to facilitate assembly, sound an alarm, or communicate the richness of a food source.

CATCH A GLIMPSE

The leafcutter ant exhibit provides an inside look at the workings of the colony. Watch them carry a piece of leaf underground and hand it off to another ant. The bit is then carefully cleaned, fertilized, and worked into the fungus garden. These ants feed on the fungi they farm, not the leaves.

Leafcutter ants use their special jaw to "saw" off pieces of plants that they carry back to their underground home. The leaves then grow a fungus that they eat.

Hummingbird Aviary

The Hummingbird Aviary is a longtime favorite of Zoo guests, a place where colorful little creatures flit by while surrounded by waterfalls and exotic plants. In addition to hovering hummingbirds, a variety of tropical birds fills the air and ground with color. The birds nest, feed, and bathe—showering several times a day. Sometimes they sit in shallow water and splash, and other times they perch next to the falls, flapping their wings and ruffling their feathers in the spray.

DID YOU KNOW?

Hummingbirds have an extremely high metabolism. They consume 77 times as much as most humans eat. That's the human equivalent of 155,000 calories a day!

Of course, hummingbirds are most known for the speed at which they beat their wings—on average, 80 times per second. Unlike other birds, they use both the upstroke and downstroke to power their flight. And since they are so skillful in flight, hummingbirds have few predators. Their biggest threat is other birds, such as jays and toucans, and the arboreal reptiles that eat their eggs and nestlings.

Coral Tree

There are more than 100 species of *Erythrina* worldwide, and the San Diego Zoo is home to 60 of them—the largest collection in the United States. Even before guests step through the entrance, they receive a blazing welcome from the Zoo's coral trees. The striking trees grace the parking lot, the entrance plaza, and much of the grounds. There's at least one coral tree or another in bloom practically year round. Many of the trees are among the largest of their species in the United States, including the *Erythrina lysistemon* by the Skyfari's eastern station.

Coral tree blossoms produce almost a thimbleful of nectar per flower, which is a great draw to the many native hummingbirds, as well as other local bird species and bees. The blossoms come in shades from pale peach to brilliant orange-red. The plants themselves also vary in appearance, from massive trees to small, sparsely branched plants.

Mission Conservation

The Zoo is collaborating with institutions in Hawaii, Southern California, and South Africa to propagate rare and threatened coral tree species and distribute them among its partners. This will allow other botanic gardens to replenish the species through seeds, cuttings, or tissue if a natural disaster occurs. Over the last year, the Zoo's Horticulture Department has worked with the Zoo's Institute for Conservation Research to look at the DNA of coral trees. Leaf samples were taken from the Zoo's trees as well as from specimens at botanical gardens in Hawaii. Researchers will gather information about the trees through DNA sequencing that can be used to identify unknown species.

Mission Conservation

– California Condor –

Since 2000, San Diego Zoo Global has operated the Hawaiian Endangered Bird Conservation Program, which has hatched and released more than 400 nene, or Hawaiian geese.

Mission Conservation

San Diego Zoo Global is committed to saving species worldwide by uniting its expertise in animal care and conservation science with its dedication to inspiring passion for nature. It is the largest zoological membership association in the world, with more than 250,000 member households and 130,000 child memberships, representing more than half a million people. This not-for-profit organization operates the San Diego Zoo, the San Diego Zoo Safari Park, and San Diego Zoo Institute for Conservation Research.

Thirty-three percent of the world's amphibians are at risk of extinction. They are threatened by chytrid fungus and habitat encroachment.

Ending Extinction

The San Diego Zoo Institute for Conservation Research, founded in 1975, is one of the largest zoo-based research centers in the world. Here, staff members generate, share, and apply scientific knowledge vital to the conservation of animals, plants, and habitats worldwide. In addition to on-site research at the San Diego Zoo and the Safari Park, the Institute conducts laboratory work at the Arnold and Mabel Beckman Center for Conservation Research and fieldwork on over 100 different conservation projects in more than 35 countries.

In 2012, the San Diego Zoo Global Wildlife Conservancy was launched to expand the organization's conservation reach. The Wildlife Conservancy helps to preserve imperiled wildlife, bank crucial genetic resources, prevent wildlife disease, conserve habitats, restore nature, and inspire change. It makes possible on-site wildlife conservation efforts at the San Diego Zoo, San Diego Zoo Safari Park, and the San Diego Zoo Institute for Conservation Research, as well as around the world.

By bringing the worldwide conservation efforts of San Diego Zoo Global under a single umbrella, the Wildlife Conservancy provides people around the world with an easy way to support the organization's efforts to save critically endangered species. People can become involved in the San Diego Zoo Global Wildlife Conservancy by joining as Wildlife Heroes or Wildlife Guardians. Both levels allow people to contribute to wildlife conservation while learning more about these species, the challenges that face them, and the work of the Wildlife Conservancy to save them. For more information about the San Diego Zoo Global Wildlife Conservancy, visit endextinction.org.

Some of the Zoo's biggest conservation success stories have been working to preserve the population of the California condor, partnering with the government of China to breed and research giant pandas, and collaborating with Australian researchers to protect koalas.

Clapper rails are bred at the San Diego Zoo and then released into protected marsh areas in Southern California.

A California condor that was hatched and raised at one of the Zoo's facilities is released into the wild. Condors are located in the wild in isolated areas of reintroduction: California, Arizona, and Mexico.

San Diego Zoo Global's work with California condors is one of its most celebrated conservation successes. In the 1980s, California condors nearly became extinct, prompting the San Diego Zoo and other groups to join forces to save the species. In 1982, only 22 birds remained in the wild. The San Diego Zoo was given permission to begin the first zoological propagation program for California condors, in cooperation with the US Fish and Wildlife Service, California Department of Fish and Wildlife, the National Audubon Society, and the Los Angeles Zoo. In 1988, the first chick bred in a zoo hatched at the Safari Park, and in 1992, the first zoo-bred condors were reintroduced to the wild in California. As of 2011, the population of condors had grown to over 400 birds, including almost 200 living in the wild.

The Zoo continues to breed condors at the San Diego Zoo Safari Park and has released more than 100 of them into the wild. In addition, researchers at the San Diego Zoo Institute for Conservation Research are spearheading a binational program to reintroduce condors into a key portion of their former range in Baja California, Mexico. By studying this population, scientists are discovering more about how condors use their habitat and find food, and how many condors can live in one area. They are also trying to find ways to keep condors safe from wind power development and to discourage condor parents from feeding their chicks microtrash—small bits of broken plastic, glass, and metal.

Koalas are one of Australia's most recognizable symbols, but their survival hangs in the balance due to loss of habitat and disease. Despite their well-loved reputation, there's still much that researchers don't know about koalas—and that can create big challenges for koala conservation and management. To shore up remaining populations, researchers from San Diego Zoo Global have teamed with Australian scientists to better understand koala mating strategies, habitat use, disease, genetics, and movement patterns. Through technology, the team has learned the importance of male bellow calls, habitat use, and scent gland communication as they relate to mating success.

The Zoo has also been investigating the seasonality of koala births on St. Bees, an island off the eastern coast of Australia, and applying its knowledge to other sites. In doing so, it has discovered a significant relationship between koala births and rainfall in central Queensland, Australia. This finding highlighted key factors needed to protect koala habitats and plan for the long-term future of wild populations. The St. Bees population of koalas has been studied for over 10 years, and the results of the studies are being compared to koalas in other sites throughout the species' home range. This work will play a role in further protecting one of Australia's most beloved, iconic animals.

At the Zoo, keepers and researchers are studying the koala colony to examine male traits, such as scent and sound, to determine the effects that they may have on female mate choice and reproduction. In addition, since 1983, the San Diego Zoo has shared koalas with zoos around the world through its Koala Loan Program. Koalas in the wild benefit from the loan program, too; funds from this program are donated to koala habitat conservation in Australia.

The Zoo conducts koala research on St. Bees Island in Australia, studying the colony's habitat use. Koalas there are fitted with GPS collars to help study their movements.

Panda cubs at the San Diego Zoo undergo a series of veterinary exams. The results are compared to the other cubs born at the Zoo, which helps researchers better understand giant panda biology.

Although giant pandas once roamed a large portion of Asia, today scientists estimate there are only 1,600 living in six ranges in southwestern China's mountains. An additional 300 pandas live in Chinese breeding facilities and zoos or are on loan to zoos around the world.

Giant pandas are considered an "umbrella species." By protecting and caring for pandas, conservationists and panda experts are also caring for other species that share their habitat. Back in the 1990s, however, biologists didn't know if they could save pandas from extinction. Little was known of their behavior in the wild, and pandas did not reproduce well in zoos. In 1996, San Diego Zoo Global partnered with Chinese colleagues at panda preserves to create a conservation strategy, which included developing early detection pregnancy tests as well as a milk formula for panda cubs that raised survival rates from zero to 100 percent.

At the Zoo, researchers collect two or more hours of data on the pandas' behavior several days a week, and they also collect physiological data. Through the pandas in the Zoo's care, it has learned much about panda husbandry, veterinary care, and nutrition; panda reproduction; the importance of environmental enrichment; and the significance of chemical communication. In 2007, San Diego Zoo Global was given permission by the Chinese government to use GPS technology to track pandas in the Foping Nature Reserve. Through this research, scientists have documented the critical importance of old growth forest in providing suitable dens for mother pandas to successfully raise their young. All this adds to our understanding of what these animals need to survive in the wild.

About the Safari Park

– African Lion –

Located near the Lion Camp in the Safari Park, Shiley's Cheetah Run shows guests the speed of the cheetah as it runs 70 miles per hour in pursuit of a mechanical lure.

About the Safari Park

Located 35 miles north of the San Diego Zoo, the San Diego Zoo Safari Park is an expansive wildlife sanctuary. Here, trails and vistas reveal memorable views of lions lounging in the grass, antelope and giraffes mingling on the savannas, and rhinos wallowing in water holes. As one of the most distinctive and successful wildlife preserves in the world, the Safari Park is a unique place where guests are immersed in the natural world and make powerful and lasting connections with wildlife. Visitors can choose from a variety of safari options, from a tram ride around the enclosures to a zip-line ride over rhinos, deer, and other animals. Many of the safaris also provide opportunities for up-close animal encounters.

Crowned cranes inhabit the Park's African Plains. In the wild, their long legs, neck, and peripheral vision help them spot predators.

The Safari Park was the brainchild of Charles Schroeder, DVM, director of the San Diego Zoo from 1953 to 1973. Envisioning a "zoo of the future" with greater space for the animals, he worked to create a haven for wild animals in zoos and to raise awareness about the crisis of vanishing species and the importance of the conservation efforts necessary to protect them. More than 18,000 mammals and 8,000 birds have been born and hatched at the Park since it opened.

The Park was originally conceived as a "back country animal preserve" that would be dedicated to breeding and maintaining species so that animals would not have to be collected from the wild. As the concept developed, it became clear that the Park could be both a preserve and a site for visitors to enjoy. More than 15 years in the making, the Safari Park was officially dedicated on May 9, 1972.

Today, the Park has more than 3,500 animals representing 400 species—many of which roam fenced habitats 100 acres in size. Its renowned botanical collection includes 3,500 species and 1.75 million specimens. Nearly half the Park's 1,800 acres has been set aside as a native species habitat.

A group of rhinos is sometimes called a "crash." The Park has the largest crash of rhinos and the most successful captive breeding program for rhinos in the world.

The Caravan Safari takes guests onto the animals' turf, enabling them to feed giraffes and rhinos and get up-close pictures of wildlife.

The 100-acre San Diego Zoo is dedicated to the conservation of endangered species and their habitats. The organization focuses on conservation and research work around the globe, educates millions of individuals a year about wildlife, and maintains accredited horticultural, animal, library, and photo collections. The Zoo also manages the 1,800-acre San Diego Zoo Safari Park (historically referred to as the Wild Animal Park), which includes an 800-acre native species reserve, and the San Diego Zoo Institute for Conservation Research. The important conservation and science work of these entities is supported in part by the Foundation of San Diego Zoo Global.

ISBN: 978-1-935442-39-4

10 9 8 7 6 5 4 3 2

San Diego Zoo: Official Guidebook was developed by Beckon Books in cooperation with the San Diego Zoo. Beckon develops and publishes custom books for leading cultural attractions, corporations, and nonprofit organizations. Beckon Books is an imprint of Southwestern Publishing Group, Inc., 2451 Atrium Way, Nashville, TN 37214. Southwestern Publishing Group, Inc., is a wholly owned subsidiary of Southwestern, Inc., Nashville, Tennessee.

Christopher G. Capen: *President and Publisher*
Betsy Holt: *Development Director*
Monika Stout: *Senior Art Director*
Kristin Connelly: *Managing Editor*
Jennifer Benson: *Proofreader*

www.beckonbooks.com
877-311-0155

Printed in the United States of America